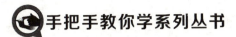 手把手教你学系列丛书

手把手教你学 FPGA

阿 东 编著

北京航空航天大学出版社

内容简介

本书主要讲解 FPGA 的程序设计,以一款热销的 FPGA 开发板为例,介绍学习 FPGA 和使用 Verilog,以及 FPGA 开发板的硬件配置,重点是第 3 章的 16 个典型实例程序,由简单到复杂,最后是 FPGA 的设计心得。

本书适合电子、通信、自动化等相关专业的本科生以及从事 FPGA 开发/IC 设计/PCB 等相关职业的初学者阅读参考。

图书在版编目(CIP)数据

手把手教你学 FPGA / 阿东编著. -- 北京:北京航空航天大学出版社,2017.2
ISBN 978-7-5124-1049-7

Ⅰ. ①手… Ⅱ. ①阿… Ⅲ. ①可编程序逻辑器件—研究 Ⅳ. ①TP332.1

中国版本图书馆 CIP 数据核字(2017)第 002529 号

版权所有,侵权必究。

手把手教你学 FPGA
阿 东 编著
责任编辑 张冀青

*

北京航空航天大学出版社出版发行

北京市海淀区学院路 37 号(邮编 100191) http://www.buaapress.com.cn
发行部电话:(010)82317024 传真:(010)82328026
读者信箱: emsbook@buaacm.com.cn 邮购电话:(010)82316936
北京市同江印刷有限公司印装 各地书店经销

*

开本:710×1 000 1/16 印张:12 字数:256 千字
2017 年 3 月第 1 版 2017 年 3 月第 1 次印刷 印数:3 000 册
ISBN 978-7-5124-1049-7 定价:35.00 元

若本书有倒页、脱页、缺页等印装质量问题,请与本社发行部联系调换。联系电话:(010)82317024

前　　言

　　本书写作的目的是让更多的同学踏入 FPGA 的世界,消除 FPGA 带给大家的一种很难学习的感觉。

　　很多同学对于学习单片机、ARM、FPGA、DSP 非常困惑,不清楚自己要学习什么,学习哪个更有前途。

　　当前,学习单片机和 ARM 的人最多,而学习 FPGA 和 DSP 的人相对少一些;单片机和 ARM 的市场应用也是最多的,基本上涉及各个领域,而 DSP 一般应用在算法领域,FPGA 一般应用在高性能处理、实时要求高的领域,比如高速接口、报文转发、图像处理、视频传输等,还可以应用在芯片前期验证。

　　如果您对算法不是很敏感,那么可以排除 DSP,笔者就对算法不是很敏感。如果您只是想找个工作,对于是单片机还是 FPGA 没有要求,那么可以好好学习一下单片机和 ARM。当然,对 ARM 的深入学习也是比较困难的,特别是要达到操作系统的级别。FPGA 本身的特性决定了其适合做高速和大容量处理,而且 FPGA 工程的方案设计相对而言比较复杂和系统化,所以学习之后,可以使你的系统观念更强,几年之后,与做单片机相比,其差异也会越来越大,成为系统工程师也就指日可待了。这也是学习 FPGA 能给您带来的一个好处,当您系统观念越来越强的时候,承担的工作也就越来越重要,待遇自然也不用多说了。

　　学习 FPGA 还有一个好处就是可以转型做芯片设计,因为做芯片设计是很多人心中的一个梦想,想当年笔者就是怀揣着这个梦想走下来的。

　　笔者在 FPGA 领域里面已经奋斗了 8 年,希望自己的一些经验,能够让学习它的人少走一些弯路。其实,FPGA 本身只是一个器件,无论是 ALTERA 的 FPGA 还是 XILINX 的 FPGA,目的都是学会使用 FPGA,而使用 FPGA 是比较简单的。

　　FPGA 的难点是逻辑设计。逻辑设计是需要长时间积累和锻炼的,不是一蹴而就的。初期最好有一套入门级的 FPGA 开发板和配套书籍,根据配套书中的实例一个一个学习,踏踏实实,多去思考,才能真正掌握逻辑设计的精髓。简单实际的例子学会了,再设计复杂点的小系统,其中的原理都是相通的。比如学会了写 100 行代码,然后再试着写 1 000 行代码,到了会写 1 000 行代码的时候,上万行甚至几十万行的代码也就不是什么难事了。本书就是基于这个思路进行编写的,由简单的流水灯开始,逐渐向复杂的 SDRAM 控制器进发。

　　现在的逻辑芯片设计,一般都是几十万行甚至上百万行的代码,大多是由几十人组成一个团队进行开发,设计和验证的工作由不同的同事分开进行,这就需要彼此默契配合和沟通。因此,同学们在学习的时候也要多讨论和交流,养成团队开发的好习惯。

另外,编程规范也需要同学们多重视,一个好的风格的代码可以节省自己和他人大量的时间,尤其是在项目开发里面,因为你的代码是需要其他人检视的。好的风格的代码一般具有清晰的注释、对齐的格式、命名能体现信号含义,逻辑清晰易懂,组合逻辑不能太复杂等。

逻辑设计能力的提高和修行差不多,都是苦差事,但是吃得苦中苦,方为人上人。从助理工程师、普通工程师、资深工程师、系统工程师到架构师,是一个漫长的修练过程。

本书是初次出版,肯定还存在这样那样的问题,希望大家反馈和指正(沟通QQ:1530384236),一起让这本书完善起来,让更多的同学加入到学习FPGA的世界!

阿　东

2016年10月8日

目 录

第1章 FPGA 概述 ... 1
1.1 为什么要学习 FPGA ... 1
1.2 学习 FPGA 的几个疑点 ... 2
1.2.1 选择 VHDL 还是 Verilog ... 2
1.2.2 NIOS 重要还是 Verilog 重要 ... 2
1.3 FPGA 简介 ... 3
1.4 Verilog 简介 ... 7
1.4.1 端口定义 ... 8
1.4.2 信号类型定义 ... 9
1.4.3 数字定义 ... 9
1.4.4 阻塞赋值和非阻塞赋值 ... 10
1.5 FPGA 开发流程 ... 11

第2章 FPGA 开发板 ... 13
2.1 STORM IV_E6 开发板简介 ... 13
2.2 STORM IV_E6 开发板详细配置 ... 15
2.3 STORM IV_E6 开发板硬件原理图 ... 16

第3章 设计实例 ... 27
3.1 LED 流水灯实验 ... 27
3.1.1 LED 简介 ... 27
3.1.2 实验任务 ... 28
3.1.3 硬件设计 ... 28
3.1.4 程序设计 ... 28
3.1.5 实验现象 ... 31
3.2 按键控制 LED 实验 ... 31
3.2.1 按键控制 LED 简介 ... 31
3.2.2 实验任务 ... 32
3.2.3 硬件设计 ... 32
3.2.4 程序设计 ... 33
3.2.5 实验现象 ... 33

- 3.3 七段数码管静态显示实验 ... 34
  - 3.3.1 数码管简介 ... 34
  - 3.3.2 实验任务 ... 36
  - 3.3.3 硬件设计 ... 36
  - 3.3.4 程序设计 ... 37
  - 3.3.5 实验现象 ... 43
- 3.4 七段数码管动态扫描实验 ... 43
  - 3.4.1 动态扫描简介 ... 43
  - 3.4.2 实验任务 ... 44
  - 3.4.3 硬件设计 ... 44
  - 3.4.4 程序设计 ... 44
  - 3.4.5 实验现象 ... 50
- 3.5 串口发送实验 ... 51
  - 3.5.1 串口简介 ... 51
  - 3.5.2 实验任务 ... 53
  - 3.5.3 硬件设计 ... 53
  - 3.5.4 程序设计 ... 53
  - 3.5.5 实验现象 ... 58
- 3.6 串口接收实验 ... 59
  - 3.6.1 串口接收简介 ... 59
  - 3.6.2 实验任务 ... 59
  - 3.6.3 硬件设计 ... 59
  - 3.6.4 程序设计 ... 60
  - 3.6.5 实验现象 ... 64
- 3.7 同步 FIFO 实验 ... 65
  - 3.7.1 同步 FIFO 简介 ... 65
  - 3.7.2 实验任务 ... 70
  - 3.7.3 硬件设计 ... 70
  - 3.7.4 程序设计 ... 70
  - 3.7.5 实验现象 ... 73
- 3.8 异步 FIFO 实验 ... 74
  - 3.8.1 异步 FIFO 简介 ... 74
  - 3.8.2 实验任务 ... 77
  - 3.8.3 硬件设计 ... 79
  - 3.8.4 程序设计 ... 79
  - 3.8.5 实验现象 ... 86

3.9 状态机实验 ··· 86
3.9.1 状态机简介 ··· 86
3.9.2 实验任务 ··· 88
3.9.3 硬件设计 ··· 88
3.9.4 程序设计 ··· 88
3.9.5 实验现象 ··· 91

3.10 EEPROM 写操作实验 ·· 91
3.10.1 EEPROM 写操作简介 ·· 91
3.10.2 实验任务 ·· 92
3.10.3 硬件设计 ·· 92
3.10.4 程序设计 ·· 92
3.10.5 实验现象 ·· 106

3.11 EEPROM 读操作实验 ·· 107
3.11.1 EEPROM 读操作简介 ·· 107
3.11.2 实验任务 ·· 107
3.11.3 硬件设计 ·· 107
3.11.4 程序设计 ·· 107
3.11.5 实验现象 ·· 119

3.12 PS/2 键盘读操作实验 ··· 120
3.12.1 PS/2 接口简介 ··· 120
3.12.2 实验任务 ·· 121
3.12.3 硬件设计 ·· 121
3.12.4 程序设计 ·· 122
3.12.5 实验现象 ·· 125

3.13 VGA 实验 ·· 126
3.13.1 VGA 简介 ·· 126
3.13.2 实验任务 ·· 129
3.13.3 硬件设计 ·· 129
3.13.4 程序设计 ·· 129
3.13.5 实验现象 ·· 132

3.14 LCD1602 实验 ·· 132
3.14.1 LCD1602 简介 ··· 132
3.14.2 实验任务 ·· 135
3.14.3 硬件设计 ·· 135
3.14.4 程序设计 ·· 136
3.14.5 实验现象 ·· 141

- 3.15 红外遥控实验 ··· 141
  - 3.15.1 红外遥控简介 ·· 141
  - 3.15.2 实验任务 ·· 144
  - 3.15.3 硬件设计 ·· 144
  - 3.15.4 程序设计 ·· 145
  - 3.15.5 实验现象 ·· 151
- 3.16 SDRAM 控制器实验 ·· 151
  - 3.16.1 SDRAM 简介 ·· 151
  - 3.16.2 实验任务 ·· 154
  - 3.16.3 硬件设计 ·· 155
  - 3.16.4 程序设计 ·· 156
  - 3.16.5 实验现象 ·· 176

第 4 章 设计思想和感悟 ··· 177
- 4.1 代码简单化 ·· 177
- 4.2 注释层次化 ·· 178
- 4.3 交互界面清晰化 ··· 178
- 4.4 模块划分最优化 ··· 178
- 4.5 方案精细化 ·· 179
- 4.6 时序流水化 ·· 179

参考文献 ·· 181

第 1 章

FPGA 概述

FPGA 的出现可以说是划时代的事件。FPGA 改变了单板硬件的设计方式,赋予 RTL 设计极大的灵活性,让 ASIC 前期物理验证有了保证,让各种高速接口有了载体,让规范协议的变化不再和硬件强相关,等等。

1.1 为什么要学习 FPGA

很多在校的学生对于学单片机、ARM、FPGA、DSP 非常困惑,不清楚自己要学习什么。当前学习单片机和 ARM 的人是最多的,而学习 FPGA 和 DSP 的人相对少一些;单片机和 ARM 的市场应用也是最多的,基本上涉及各个领域,而 DSP 一般应用在算法领域,FPGA 一般应用在高性能处理、实时要求高的领域,比如高速接口、报文转发、图像处理、视频传输等,还可以应用在芯片前期验证。

如果你对算法不是很熟悉,那么可以排除 DSP。如果你只是想找个工作,对于是单片机还是 FPGA 没有要求,那么可以好好学习一下单片机和 ARM。当然,对 ARM 的深入学习也还是比较困难的,特别是要达到操作系统级别。FPGA 本身的特性决定了它适合做高速和大容量处理,而且 FPGA 工程的方案设计相对比较复杂和系统化,所以学习它之后,你的系统观念会更强,几年之后,与做单片机的差异也会越来越大,很可能那时你就是系统工程师了。这也是学习 FPGA 所能带来的一个好处,当你的系统观念越来越强时,承担的工作也就越来越重要,待遇自然也会越来越好。

学习 FPGA 还有一个好处就是可以转型做芯片设计,因为做芯片设计是很多人心中的一个梦想,想当年笔者就是怀揣着这个梦想走下来的。

1.2 学习 FPGA 的几个疑点

1.2.1 选择 VHDL 还是 Verilog

VHDL 和 Verilog 都是用于数字电子系统设计的硬件描述语言,而且两种语言都是 IEEE 的标准。

VHDL 诞生于 1982 年。在 1987 年底,VHDL 被 IEEE 和美国国防部确认为标准硬件描述语言。自 IEEE 公布了 VHDL 的标准版本——IEEE-1076(简称 87 版)之后,各 EDA 公司相继推出了自己的 VHDL 设计环境,或宣布自己的设计工具可以和 VHDL 接口。此后,VHDL 在电子设计领域被广泛接受,并逐步取代了原有的非标准的硬件描述语言。而 Verilog HDL 是由 GDA(Gateway Design Automation)公司的 Phil Moorby 在 1983 年末首创的,最初只设计了一个仿真与验证工具,之后又陆续开发了相关的故障模拟与时序分析工具。1985 年,GDA 推出它的第三个商用仿真器 Verilog-XL,获得了巨大的成功,从而 Verilog HDL 得到迅速推广应用。1989 年,CADENCE 公司收购了 GDA 公司,Verilog HDL 成为了该公司的独家专利。1990 年,CADENCE 公司公开发表了 Verilog HDL,并成立了 LVI 组织,以促进 Verilog HDL 成为 IEEE 标准,即 IEEE Standard 1995。

2001 年,IEEE 发布了 Verilog—2001 标准。Verilog—2001 相对于 Verilog—1995 做了很多有用的改进,给编程人员带来很大的帮助。

很多初学者对于学习 FPGA 是使用 VHDL 还是 Verilog 没有方向,很多学校目前还是在教 VHDL,真是误人子弟。之前,一位初学者学习了一个月的 VHDL,然后才问我该学习 VHDL 还是 Verilog。在此特别强调一下,现在基本上所有的芯片和设计都采用 Verilog,本人也经历了几个大型项目,全部采用 Verilog,包括 ARM 的芯片和 IP 都是采用 Verilog,使用 VHDL 的人非常非常少,只有部分学校还在教学和少数研究所在用。Verilog 已经占据了完全主导的地位。

它们的设计思想都是一样的,为什么还要学习大公司都不用的东西呢?

1.2.2 NIOS 重要还是 Verilog 重要

NIOS 嵌入式处理器是 Altera 公司推出的采用哈佛结构、具有 32 位指令集的第二代片上可编程的软核处理器,其最大优势和特点是模块化的硬件结构,以及由此带来的灵活性和可裁剪性。

因此很多初学者可能觉得 NIOS 很重要,但是很遗憾地告诉大家,FPGA 的主流和优势设计是使用 Verilog。为什么不是 NIOS 呢?因为 NIOS 使用 C 语言开发,需要在 FPGA 宝贵的逻辑资源内建一个 CPU,速度可能还不及一般的 ARM 芯片。另外,功耗、成本都比 ARM 高很多,易用性也比 ARM 差很多。在一些 ARM 满足不了的场景使用 NIOS 倒是不错,比如系统需要 3 个以上串口等多个相关外设,或者单板上已经有大容量的 FPGA(FPGA 资源不使用也浪费)。

FPGA 为什么还那么流行呢? 因为 FPGA 是逻辑器件,内部都是由可以编程的寄存器、组合逻辑构成,使用硬件语言编程实现,可以一个时钟周期执行一次动作,甚至 0 个时钟周期都可以执行操作(完全组合逻辑实现),而单片机/ARM 需要 CPU 取指、译码、执行,自然 FPGA 的速度非常快,这一点是 ARM 替代不了的。

1.3　FPGA 简介

FPGA(Field-Programmable Gate Array),即现场可编程门阵列,它是在 PAL(Programmable Array Logic,可编程阵列逻辑)、GAL(Generic Array Logic,通用可编程阵列逻辑)、CPLD(Complex Programmable Logic Device,复杂的可编程逻辑器件)等可编程器件的基础上进一步发展的产物。它是作为专用集成电路(ASIC)领域中的一种半定制电路而出现的,既解决了定制电路的不足,又克服了原有可编程器件门电路数有限的缺点。

可以说,FPGA 是划时代的发明,解决了硬件逻辑可以任意编程而不需要改动硬件的问题,提高了硬件高速处理的能力,同时给 ASIC 前端验证提供了非常重要的手段。

图 1.1 是 FPGA 的结构原理图,内部包含了很多个物理单元。

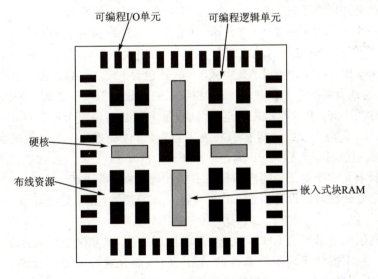

图 1.1　FPGA 结构原理图

下面介绍每个单元的基本概念。

1. 可编程输入/输出单元

输入/输出单元简称 I/O 单元。它们是芯片与外界电路的接口部分,用于完成不同电气特性下对输入/输出信号的驱动与匹配需求。为了使 FPGA 有更灵活的应用,目前大多数 FPGA 的 I/O 单元被设计为可编程模式,即通过软件的灵活配置,可以适配不同的电气标准与 I/O 物理特性;可以调整匹配阻抗特性,上下拉电阻;可以调整驱动电流的大小等。

可编程 I/O 单元支持的电气标准因工艺不同而异,因为不同器件商、不同器件族的 FPGA 支持的 I/O 标准不同。一般来说,常见的电气标准有 LVTTL、LVCMOS、SSTL、HSTL、LVDS、LVPECL、PCI 等。值得一提的是,随着 ASIC 工艺的飞速发展,目前可编程 I/O 接口支持的最高频率越来越高,一些高端 FPGA 通过 DDR 寄存器技术,甚至可以支持高达 2 Gbit/s 的数据速率。

2. 基本可编程逻辑单元

基本可编程逻辑单元是可编程逻辑的主体,可以根据设计灵活地改变其内部连接与配置,完成不同的逻辑功能。FPGA 一般是基于 SRAM 工艺的,其基本可编程逻辑单元几乎都是由查找表和寄存器组成的。FPGA 内部查找表一般分为 4 个输入,查找表完成纯组合逻辑功能。FPGA 内部寄存器结构相当灵活,可以配置为带同步/异步复位或置位、时钟使能的触发器,也可以配置成锁存器。FPGA 一般依赖寄存器完成同步时序逻辑设计。比较经典的基本可编程逻辑单元的配置是一个寄存器加一个查找表,但是不同的厂商,寄存器与查找表也有一定的差异,而且寄存器与查找表的组合模式也会不同。例如,Altera 可编程逻辑单元通常称为 LE(Logic Element),由一个寄存器加一个 LUT(Look-Up-Table,显示查找表)构成。Altera 的大多数 FPGA 将 10 个 LE 有机地组合在一起,构成更大的逻辑单元——逻辑阵列模块(Logic Array Black,LAB)。LAB 中除了 LE 还包含 LE 之间的进位链、LAB 控制信号、局部互连线资源、LUT 级联链、寄存器级联链等连线与控制资源。Xilinx 可编程逻辑单元称为 Slice,由上、下两个部分组成,每个部分都由一个寄存器加一个 LUT 组成,称为 LC。两个 LC 之间有一些公用逻辑,可以完成 LC 之间的配合与级联。Lattice 的底部逻辑单元称为 PFU,由 8 个 LUT 和 8~9 个寄存器构成。当然,这些可编程逻辑单元的配置结构随着器件的不断发展也在不断更新,最新的一些可编程逻辑器件常常根据需求设计新的 LUT 和寄存器的配置比率,并优化其内部的连接构造。

学习底层配置单元的 LUT 和寄存器的配置比率,其重要意义在于器件选型和规模估算。很多器件手册上用器件的 ASIC 门数或等效的系统门数表示器件的规模。但是,由于目前 FPGA 内部除了基本可编程逻辑单元外,还包含丰富的嵌入式 RAM、PLL 或 DLL,以及专用的 Hard IP Core(如 PCIE、Serdes 硬核)等。这些功能模块也会等效出一定规模的系统门,所以用系统门权衡基本可编程逻辑单元的数量是不准确的,常常让设计者混淆。比较简单而科学的方法是用器件的寄存器或 LUT 的数量衡量。例如,Xilinx 的 Spartan 系列 XC3S1000 有 15 360 个 LUT,而 Lattice 的 EC 系列 LFEC15E 也有 15 360 个 LUT,所以这两款 FPGA 的可编程逻辑单元数量相当,属于同一规模的产品。Altera 的 Cyclone Ⅳ 器件族 EP4CE6 的 LUT 数量是 6 000 个,就比前面提到的两款 FPGA 规模略小。需要说明的是,器件选型是一个综合性的问题,需要综合考虑设计的需求、成本、规模、速度等级、时钟资源、I/O 特性、封装、专用功能模块等诸多因素。

3. 嵌入式块 RAM

目前,大多数 FPGA 都有内嵌的块 RAM。FPGA 内部嵌入可编程 RAM 模块,大大地拓展了 FPGA 的应用范围和使用灵活性。FPGA 内嵌的块 RAM 一般可配置为单端口 RAM、双端口 RAM、伪双口 RAM、内容可寻址存储器(Coment-Addressable Memory,CAM)、FIFO 等常用存储结构。RAM 的概念和功能,读者应该非常熟悉,在此不再赘述。FPGA 中其实并没有专用的 ROM 硬件资源,实现 ROM 的思路是对 RAM 赋初值。所谓 CAM,也称内容地址存储器,CAM 这种存储器在其每个存储单元都包含了一个比较的逻辑,写入 CAM 的数据会与其内部存储的每一个数据进行比较,并返回与端口数据相同的所有内部数据的地址。概括地讲,RAM 是一种根据地址读、写数据的存储单元,而 CAM 恰恰相反,它返回的是与端口数据相同的所有内部地址。CAM 的应用十分广泛,比如在路由器中的交换表等。FIFO 是先进先出队列的存储结构。FPGA 内部实现的 RAM、ROM、CAM、FIFO 等存储结构都可以基于嵌入式块 RAM 单元,并根据需求自动生成相应的粘合逻辑以完成地址和片选等控制逻辑。

不同器件商或不同器件族的内嵌块 RAM,其结构也不同。Xilinx 常见的块 RAM 大小是 4 Kbit 和 18 Kbit,Lattice 常用的块 RAM 大小是 9 Kbit,Altera 的块 RAM 最灵活。一些高端器件内部同时含有 3 种块 RAM 结构,分别是 M512 RAM、M4K RAM、M9K RAM。

需要补充的一点是,除了块 RAM,Xilinx 和 Lattice 的 FPGA 还可以灵活地将 LUT 配置成 RAM、ROM、FIFO 等存储结构,这种技术称为分布式 RAM。根据设计需求,块 RAM 的数量和配置方式也是器件选型的一个重要标准。

4. 丰富的布线资源

布线资源连通 FPGA 内部的所有单元,而连线的长度和工艺决定着信号在连线上的驱动能力和传输速度。FPGA 芯片内部有着丰富的布线资源,根据工艺、长度、宽度和分布位置的不同而划分为 4 种不同的类别。

第一类是全局布线资源,用于芯片内部全局时钟和全局复位/置位的布线;
第二类是长线资源,用于完成芯片 Bank 间的高速信号和第二全局时钟信号的布线；

第三类是短线资源，用于完成基本逻辑单元之间的逻辑互连和布线；

第四类是分布式的布线资源，用于专有时钟、复位等控制信号线。

在实际中设计者不需要直接选择布线资源，布局布线器可自动地根据输入逻辑网表的拓扑结构和约束条件选择布线资源来连通各个模块单元。从本质上讲，布线资源的使用方法与设计的结果有密切、直接的关系。

5. 底层嵌入功能单元

底层嵌入功能单元的概念比较笼统，本文指的是那些通用程度较高的嵌入式功能模块，比如PLL、DLL、DSP、CPU等。随着FPGA的发展，这些模块被越来越多地嵌入到FPGA的内部，以满足不同场合的需求。

目前，大多数FPGA厂商都在FPGA内部集成了DLL或者PLL硬件电路，用于完成时钟的高精度、低抖动的倍频、分频、占空比调整、相移等功能。高端FPGA产品集成的DLL和PLL资源越来越丰富，功能越来越复杂，精度越来越高。Altera芯片集成的是PLL，Xilinx集成的是DLL，Lattice的新型FPGA同时集成了PLL与DLL以适应不同的需求。Altera芯片的PLL模块分为增强型PLL和快速PLL等，Xilinx芯片的DLL模块名称为CLKDLL。在高端FPGA中，CLKDLL的增强型模块为DCM。这些时钟模块的生成和配置方法一般分为两种。一种是在HDL代码和原理图中直接例化，另一种是在IP核生成器中配置相关参数，自动生成IP。Altera的IP核生成器叫做Mega wizard，Xilinx的IP核生成器叫做Core Generator，Lattice的IP核生成器叫做Module/IP manager。另外，可以通过在综合、实现步骤的约束文件中编写约束属性来完成时钟模块的约束。

越来越多的高端FPGA产品将包含DSP或CPU等软处理核，从而FPGA将由传统的硬件设计手段逐步过渡到系统级设计平台。例如Altera的Stratix、StratixⅡ、StratixⅣ等器件族内部集成了DSP Core，配合同样的逻辑资源，还可实现ARM、MIPS、NIOSⅡ等嵌入式处理系统；Xilinx的VirtesⅡ和VirtexⅡ pro系列FPGA内部集成了Power PC450的CPU Core和MicroBlaze RISC处理器Core；Lattice的ECP系列FPGA内部集成了系统DSP Core模块。这些CPU或DSP处理模块的硬件主要由一些加、乘、快速进位链、Pipelining和Mux等结构组成，用逻辑资源和块RAM实现软核部分，组成了功能强大的软运算中心。这种CPU或DSP比较适合实现FIR滤波器、编码解码、FFT等运算密集型应用。FPGA内部嵌入CPU或DSP等处理器，使FPGA在一定程度上具备了实现软硬件联合系统的能力，FPGA正逐步成为SPOC的高效设计平台。Altera的系统级开发工具有SOPC Buider、DSP Builder，以及专用硬件结构与软硬件协同处理模块等；Xilinx的系统开发工具有EDK和Platform Studio；Lattice的嵌入式DSP开发工具有Matlab的Simulink。

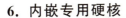

6. 内嵌专用硬核

这里的内嵌专用硬核与前面的底层嵌入单元是有区分的,这里讲的内嵌专用硬核主要指那些通用性相对较弱,不是所有 FPGA 器件都包含的硬核。我们称 FPGA 和 CPLD 为同样逻辑器件,是区别于专用集成电路而言的。其实,FPGA 内部也有两个阵营:一方面是通用性强,目标市场范围很广,价格适中的 FPGA;另一方面是针对性较强,目标市场明确,价格较高的 FPGA。前者主要指低成本 FPGA,后者主要面向某些高端通信市场的可编程逻辑器件。

1.4 Verilog 简介

Verilog 语法标准规定的内容还是很多的,可以单独写一本书了,但是本书项目里面真正用到的不多,本节不准备描述所有的语法,只选择实际项目中用到的几个语法。

阿东也不建议大家一上来就拼命学习各种语法,知道常用的几个语法就可以了,关键还在于学习实践 Verilog 的设计思想。如果项目里需要用到复杂的语法,打开相关的语法书看一下就知道了。

下面通过流水灯实验来讲解基本的语法,代码如下:

```
module LED(
                    //input
    input       sys_clk,    //system clock
    input       sys_rst_n,  //system reset,low is active
                    //output
    output      reg[7:0]LED
    );

//reg define
reg [2:0] count;
reg [24:0] counter;

//***********************************************************
//**                     Main Program
//***********************************************************

//count for add counter,0.5 s/20 ns = 25 000 000,need 25 bit cnt
always @(posedge sys_clk or negedge sys_rst_n) begin
    if ( sys_rst_n == 1'b0 )
        counter <= 25'b0;
    else
        counter <= counter + 25'b1;
```

```verilog
        end

        always @(posedge sys_clk or negedge sys_rst_n) begin
            if ( sys_rst_n == 1'b0 )
                count <= 3'b0;
            else if( counter == 25'd0 )
                count <= count + 3'b1;
            else;
        end

        //ctrl the LED pipiline display when count is equal to 0,1..
        always @(posedge sys_clk or negedge sys_rst_n) begin
            if ( sys_rst_n == 1'b0 )
                LED <= 8'b0;
            else begin
                case ( count )              //通过控制 I/O 口的高低电平实现发光二极管的亮灭
                    0: LED = 8'b00000001;
                    1: LED = 8'b00000010;
                    2: LED = 8'b00000100;
                    3: LED = 8'b00001000;
                    4: LED = 8'b00010000;
                    5: LED = 8'b00100000;
                    6: LED = 8'b01000000;
                    7: LED = 8'b10000000;
                    default: LED = 8'b00000000;
                endcase
            end
        end

endmodule
//end of rtl code
```

1.4.1 端口定义

模块的端口可以是输入端口、输出端口或双向端口。缺省的端口类型为线网类型(即 wire 类型)。输入端口默认为 wire 类型,不需要定义,输出端口或双向端口可声明为 wire/reg 类型,使用 reg 必须显式声明,使用 wire 也强烈建议显式声明。

Verilog 2001 语法中的端口定义包括端口名、信号的输入/输出、wire/reg 类型、位宽定义,极大地减少了端口声明占用的代码行数。当前 Verilog 2001 语法已经非常普及。

例子：

```
module LED (
            //input
            input sys_clk,          //system clock
            input sys_rst_n ,       //system reset, low is active
            //output
            output reg [7:0] LED );
```

该例子包括输入、输出、名称、位宽、输出的信号类型。

1.4.2 信号类型定义

信号类型主要有 wire 和 reg 两种。线网类型用于对结构化器件之间的物理连线的建模。由于线网类型代表的是物理连接线，因此它不存储逻辑值，必须由器件驱动，通常由 assign 进行赋值。例如：

assign A = B ^ C;

reg 是最常用的寄存器类型，寄存器类型通常用于对存储单元的描述，如 D 型触发器、ROM 等。当存储器类型的信号在某种触发机制下分配了一个值，在分配下一个值时应保留原值。

reg 类型定义语法如下：

reg [msb: lsb] reg1, reg2, ... reg N;

说明：

reg 类型的信号不一定是寄存器。Verilog 语法规定的 Always 语句都要使用 reg 定义，而 Always 语句分为带时钟和不带时钟两种。不带时钟的，综合出来就是组合逻辑；带时钟的，综合出来才是寄存器。

1.4.3 数字定义

[size] 'base value

其中：size 定义以位计的常量的位长。

base 为 o 或 O（表示八进制），b 或 B（表示二进制），d 或 D（表示十进制），h 或 H（表示十六进制）之一。

value 是基于 base 的值的数字序列。值 x 和 z 以及十六进制中的 a~f 不区分大小写。

下面是一些具体实例：

5'o37 表示 5 位八进制数(二进制 11111);

4'd2 表示 4 位十进制数(二进制 0011);

4'b1x_01 表示 4 位二进制数;

7'hx 表示 7 位 x(扩展的 x),即 xxxxxxx;

4'hz 表示 4 位 z(扩展的 z),即 zzzz;

4'd-4 表示非法数值,因为数值不能为负;

3' b 001 表示非法数值,001 和基数 b 之间不允许出现空格;

(2+3)'b10 表示非法数值,位长不能够为表达式。

说明:

实际项目中使用十进制、二进制和十六进制较多,八进制使用较少。

1.4.4 阻塞赋值和非阻塞赋值

1. 阻塞赋值

"="用于阻塞的赋值,凡是在组合逻辑(如在 assign 语句中)赋值的请用阻塞赋值。阻塞赋值"="在 begin 和 end 之间的语句是顺序执行的,属于串行语句。

一个组合逻辑的例子:

```
always @( * ) begin
    if ( new_vld_after == 1'b1 )
        port_win = new_port_after ;
    else if ( new_vld_before == 1'b1 )
        port_win = new_port_before ;
    else
        port_win = last_sel_port ;
end
```

注意:

always 语句的敏感变量如果不含有时钟,即 always(*),那么也属于组合逻辑,需要使用阻塞赋值。

2. 非阻塞赋值

"<="用于非阻塞的赋值,凡是在时序逻辑(如在 always 语句中)赋值的请用非阻塞赋值。非阻塞赋值"<=",在 begin 和 end 之间的语句是并行执行的,属于并行执行语句。

一个时序逻辑的例子:

```
always @(posedge sys_clk or negedge sys_rst_n) begin
    if ( sys_rst_n == 1'b0 )
```

```
            counter <= 25'b0;
        else
            counter <= counter + 25'b1;
end
```

注意：

时序逻辑值是带有时钟的 always 块逻辑，只有 always 带有时钟，这个逻辑才能综合为寄存器。

1.5 FPGA 开发流程

FPGA 的典型开发流程如图 1.2 所示。

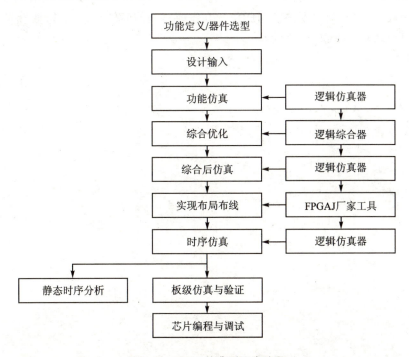

图 1.2 FPGA 的典型开发流程

在图 1.2 中，逻辑仿真器主要有 Modelsim 等，逻辑综合器主要有 Quartus Ⅱ、Synplify Pro、ISE 等。FPGA 厂家集成的开发环境有 Altera 公司的 Quartus Ⅱ，Xilinx 公司的 ISE。

设计输入主要有原理图输入和 HDL 输入两种方式，在大部分的设计中，FPGA 和 ASIC 的工程师都使用 HDL 输入。

设计仿真主要包括功能仿真和网表仿真。设计仿真需要 RTL 代码或综合后的 HDL 网表和验证程序，有时候还需要测试数据，测试数据可能是代码编译后的二进制文件或是使用专门的工具采集的数据。

布局布线工具利用综合生成的网表、调用模块的网表,根据布局布线目标,把设计翻译成原始的目标工艺,最后得到生成编程比特流所需的数据文件。布局布线目标包括所使用的 FPGA 具体型号等,约束条件包括引脚位置、引脚电平逻辑(LVTIL、LCMOS 等)需要达到的时钟频率,有时还包括部分模块的布局、块 RAM 的位置等。

静态时序分析主要是通过分析每个时序路径的延时,计算出设计的各项时序性能指标,如最高时钟频率、建立保持设计等。发现时序违规,如果工具有时序告警,有时需要检查设计,对设计进行修改,以满足时序要求,比如组合逻辑过多,导致寄存器建立时间不满足,就需要对组合逻辑中间插入一些寄存器。它仅仅涉及时序性能的分析,不涉及设计的逻辑功能。

在一般设计中,只需要注意引脚位置和需要达到的时钟频率,逻辑端口与 FPGA 引脚的对应取决于 PCB 板的设计。

第 2 章 FPGA 开发板

本章主要介绍暴风四代 FPGA 开发板(STORMIV_E6)的硬件结构,只有了解了开发板的硬件原理,才能知道如何通过程序驱动相关的外设。

2.1　STORM IV_E6 开发板简介

暴风四代开发板实物图片如图 2.1 所示。

图 2.1　暴风四代开发板实物图

暴风四代开发板功能示意图如图 2.2 所示（FPGA 为 EP4CE6）。

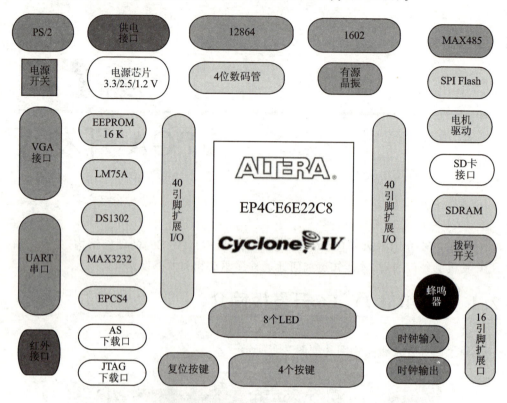

图 2.2　暴风四代开发板功能示意图

STORM IV_E6 开发板使用 Altera 公司的 Cyclone IV FPGA 芯片。四代的 FPGA 速度、功耗都比之前的 Cyclone I/II/III 系列 FPGA 优化很多。开发板的配套外设非常丰富，有 SDRAM、串行 Flash、1602、12864、LED、按键、蜂鸣器、串口、VGA 接口、扩展口、MAX485、电机驱动接口等。

这一款开发板非常适合初学者入门学习，而且开发板的价格也非常便宜，仅 100 多元，还赠送 USB Blaster 下载器，淘宝链接如下：

http://item.taobao.com/item.htm?id=35911884243

二维码如下：

2.2　STORM IV_E6 开发板详细配置

- 1 片 Cyclone IV FPGA——EP4CE6E22C8；
- 1 片 EPCS4 配置芯片，4 Mbit 容量 Flash；
- USB 接口提供电源；
- 1 个 50 MHz 有源晶振；
- 8 个发光二极管(LED)；
- 4 个通用按键；
- 1 个复位按键；
- 1 片 SDRAM，容量为 64 Mbit；
- 1 个 4 位拨码开关；
- 1 个 4 位七段数码管；
- 1 个 DS1302 时钟芯片；
- 1 个红外遥控接口；
- 1 个 MAX485 接口；
- 1 个电机接口：支持 1 路步进电机，1 路直流电机；
- 1 片 I2C 接口的 EEPROM，AT24C16，16 Kbit 容量；
- 1 片串行 Flash：16 Mbit 容量；
- 1 个 5 V 有源蜂鸣器；
- 1 个 RS232 UART 串口；
- 1 个 PS/2 接口，可以接键盘或者鼠标；
- 1 个 VGA 接口，可以接计算机显示器；
- 1 个温度传感器 LM75A；
- 1 个 LCD1602 液晶接口，引脚兼容 1.8 英寸 TFT 彩屏液晶；
- 1 个 LCD12864 液晶接口；
- 1 个专用时钟输入接口；
- 1 个专用时钟输出接口；
- 2 个 40 引脚扩展接口(排针)；
- 1 个 16 引脚扩展接口(排母)，精心设计的 16 引脚扩展口，可以直接插网络模块、无线模块等独立模块；
- 1 个 JTAG 接口；
- 1 个 AS 接口；
- 1 个 SD 卡接口

2.3 STORM IV_E6 开发板硬件原理图

暴风四代开发板有 8 个 LED 指示灯、4 个通用按键、1 个复位按键、4 个拨码开关、1 个 50 MHz 的有源晶振、1 个七段 4 位数码管、1 个 5 V 有源蜂鸣器、1 个 PS/2 接口、标准的 RS232 串口、1 个 VGA 接口、1 个 16 Kbit 的 I2C 接口的 EEPROM、1 个 I2C 接口的温度传感器 LM75A、1 个 RS485 接口、1 个 PS1302 数字时钟芯片、1 个 16 Mbit 的 SPI 接口的串行 Flash、1 个电机驱动电路、1 个时钟扩展输入口、1 个时时钟扩展输出口、1 个 JTAG 接口、1 个 AS 接口、1 个 LCD1602 接口、1 个 LCD12864 接口、1 个 SDCARD 接口、1 个 64 Mbit 的 SDRAM 芯片、1 个 4 Mbit 的 Flash 芯片 EPCS4 及 3 个扩展口（J3、J4、J8），它们的电路原理图如图 2.3～图 2.30 所示。下面对各电路原理进行详细介绍。

1. 发光二极管

暴风四代开发板有 8 个 LED 指示灯（见图 2.3）。当 FPGA 输出为 1 时 LED 发光，330 Ω 排阻（R10，R13）对 LED 灯起限流作用，8 个 LED(D1～D8)可用于 8 bit 的数据显示。

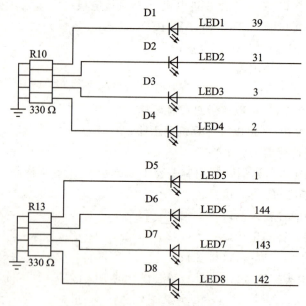

图 2.3 发光二极管原理图

2. 通用按键

暴风四代开发板有 4 个通用按键（见图 2.4）。当按键 KEY1～KEY4 没有被按下时输入为 1，当按键被按下时输入为 0；10 KΩ 排阻（R11）可以起上拉功能的作用，按键没有被按下 I/O 会被上拉到高电平。

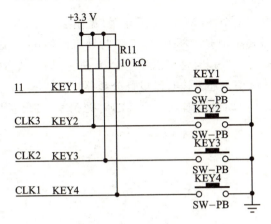

图 2.4 通用按键原理图

3. 复位按键

暴风四代开发板有 1 个复位按键(见图 2.5)。当按键 RESET 没有被按下时输入为 1,当按键 RESET 被按下时输入为 0;复位时是 0 有效(代表复位)。一般逻辑设计开发,都是低有效,即按下复位按键有效。

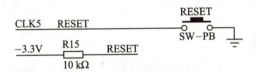

图 2.5 复位按键原理图

4. 拨码开关

暴风四代开发板有 4 个拨码开关(见图 2.6)。当拨码开关 S1 没有被拨下时输入为 1,当拨码开关 S1 被拨下时输入为 0;10 kΩ 排阻 R3 起上拉功能作用,拨码开关没有被关上 I/O 会被上拉到高电平。

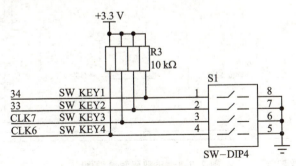

图 2.6 拨码开关原理图

5. 有源晶振

暴风四代开发板有 1 个 50 MHz 有源晶振（见图 2.7），系统所使用的时钟就是由它提供的。如果需要多个时钟频率，则高频时钟可以使用 PLL 进行倍频（或者使用时钟输入口进行输入），低频时钟可以使用 PLL 进行降频（或者使用时钟输入口进行输入自定义的时钟频率），还可以自己设计分频器。

时钟输入口（见图 2.8）可以输入自定义的时钟频率，方便扩展输入不同的时钟频率。

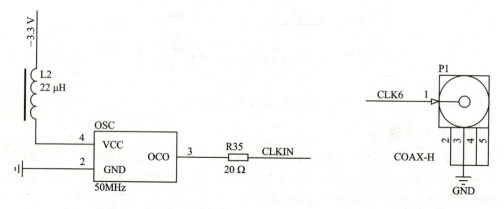

图 2.7　有源晶振原理图　　　　　　　图 2.8　时钟输入口原理图

6. 七段数码管

暴风四代开发板有一个七段 4 位数码管（见图 2.9），为共阳的数码管。也就是说，数码管显示需要片选端为 1，段位端为 0，才能导通显示。该数码管可以显示 4 个

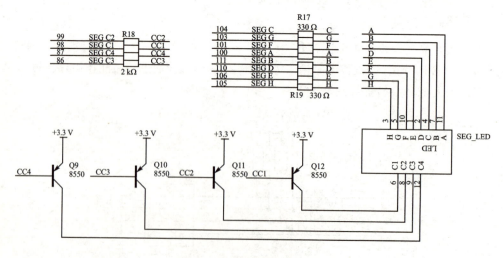

图 2.9　七段数码管原理图

数字,每个数字有不同的片选信号。为了增加驱动能力,采用 PNP 三极管 8550。可以使用静态驱动或者动态扫描方式进行数码管的驱动。

7. 蜂鸣器

暴风四代开发板有一个 5 V 有源蜂鸣器(见图 2.10)。该蜂鸣器 3.3 V 可发声,使用 FPGA I/O 直接驱动。FPGA I/O 输出高电平,蜂鸣器可发声;输出低电平,蜂鸣器不发声。该蜂鸣器是有源蜂鸣器。

8. PS/2 接口

暴风四代开发板有一个 PS/2 接口(见图 2.11)。PS/2 的两个引脚通过 VGA 的串接电阻接到 FPGA 的 I/O 上,起隔离的作用;另外,还可以和 VGA 的 HS/VS 复用引脚。该 PS/2 接口可以接 PS/2 接口的键盘或者鼠标。

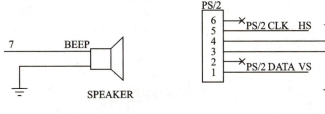

图 2.10 蜂鸣器原理图   图 2.11 PS/2 接口原理图

9. UART 串口

暴风四代开发板有标准的 RS232 串口(见图 2.12),可以作为通用串口,或者作为调试接口。串口使用 MAX3232 芯片隔离串口 TTL 电平和 FPGA CMOS 电平。

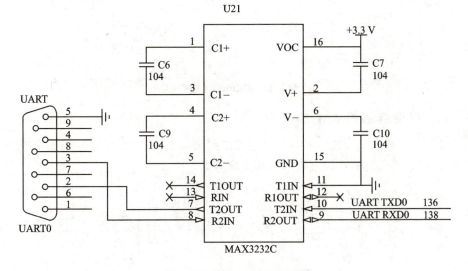

图 2.12 UART 串口原理图

10. VGA 接口

暴风四代开发板有一个 VGA 接口(见图 2.13)，使用 3 bit FPGA I/O 作为 RGB 信号线，可以产生 8 种颜色。

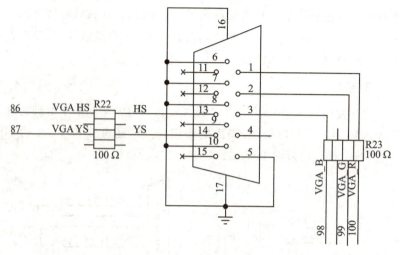

图 2.13 VGA 接口原理图

11. EEPROM

暴风四代开发板有一个 16 Kbit 的 I2C 接口的 EEPROM(见图 2.14)，可以存储数据和图片等。排阻 R12 为 EEPROM 和 LM75A 共用的上拉电阻。

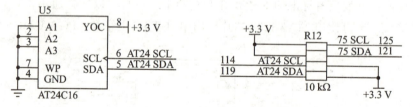

图 2.14 EEPROM 原理图

12. 温度传感器 LM75

暴风开发板有一个使用 I2C 接口的温度传感器 LM75A。该芯片内部自带 A/D 转换，可以直接通过 I2C 接口读取内部温度数据。

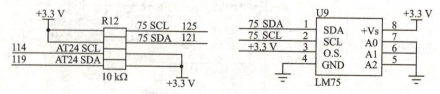

图 2.15 温度传感器原理图

13. MAX485 接口

暴风四代开发板有一个 RS485 接口(见图 2.16),通过 MAX485 芯片完成电平转换,J5 可以接 RS485 电平。

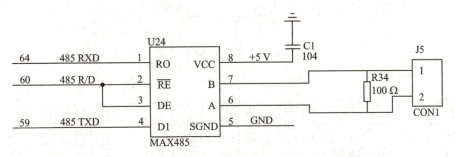

图 2.16　MAX485 原理图

14. DS1302 数字时钟

暴风四代开发板有一个 DS1302 数字时钟芯片(见图 2.17)。该芯片可以产生时间和日期,能够做时钟和万年历等。该芯片需要一个 32 kHz 的时钟和一个备用电池,备用电池为开发板断电的时候给 DS1302 芯片提供电源。如果只是调试使用,则可以不用备用电池。

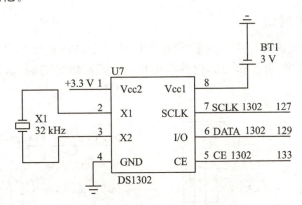

图 2.17　DS1302 数字时钟原理图

15. 串行 Flash

暴风四代开发板有一个 16 Mbit 的 SPI 接口的串行 Flash(见图 2.18)。串行 Flash 可以存储程序、数据等。

16. 电机驱动

暴风四代开发板有一个电机驱动电路(见图 2.19),使用 ULN2003 电机驱动芯片,可以驱动一个 5 V 步进电机或者一个直流电机。

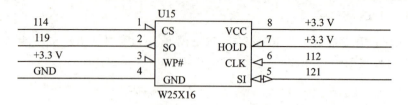

图 2.18 串行 Flash 原理图

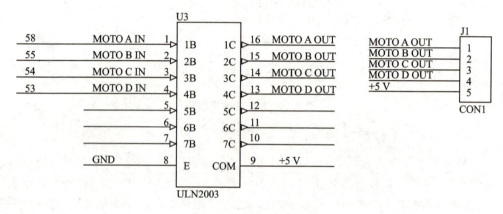

图 2.19 电机驱动原理图

17. 时钟扩展口

暴风四代开发板有一个时钟扩展输入口和一个时钟扩展输出口(见图 2.20)。P2 是时钟扩展输出,可以作为其他外设使用;P1 是时钟扩展输入,可以作为 FPGA 内部时钟使用。

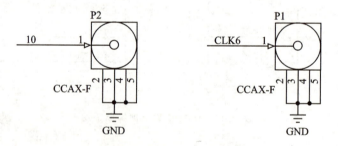

图 2.20 时钟扩展口原理图

18. JTAG 与 AS 接口

暴风四代开发板有 JTAG 接口(见图 2.21)和 AS 接口(见图 2.22)。JTAG 接口下载程序掉电会消失,程序直接下载到 FPGA;AS 接口下载程序掉电不会消失,程序直接烧写到配置 Flash。

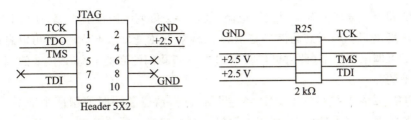

图 2.21　JTAG 接口原理图

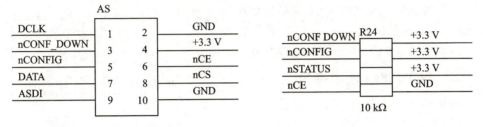

图 2.22　AS 接口原理图

19. LCD1602 液晶接口

暴风四代开发板有一个 LCD1602 接口（见图 2.23）。

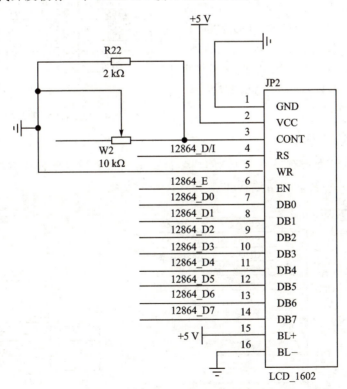

图 2.23　LCD1602 液晶接口原理图

R22 和 W2 全部为调节对比度电阻,默认 W2 可调电阻没有焊接,使用 R22 作为调节对比度电阻。如果您的液晶在 R22 的调节下还是显示不清楚,则可以自己焊接一个 W2 可调电阻调节(此时可以将 R22 取下),我们配套的液晶在 R22 的调节下是非常清楚的。

LCD1602 液晶接口和 LCD12864 共用引脚,请避免同时使用两个液晶。另外,由于液晶需要使用 5 V 电源,所以在测试液晶的时候必须给开发板的 USB 电源接 5 V 电源,否则液晶不工作。

我们提供的 1.8 寸的 TFT 液晶也是使用 LCD1602 液晶接口,1.8 寸 TFT 彩屏接口可以直接插在 LCD1602 液晶接口上面。

20. LCD12864 液晶接口

暴风四代开发板有一个 LCD12864 接口(见图 2.24)。

W3 为调节对比度电阻,默认 W3 可调电阻没有焊接,如果您的液晶显示不清楚,那么可以自己焊接一个 W3 可调电阻,我们配套的液晶在无 W3 的调节下是非常清楚的。

LCD1602 液晶接口和 LCD12864 共用引脚,请避免同时使用两个液晶。另外,由于液晶需要使用 5 V 电源,所以在测试液晶的时候必须给开发板的 USB 电源接 5 V 电源,否则液晶不工作。

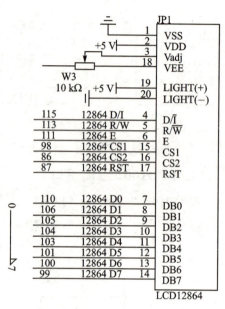

图 2.24　LCD12864 液晶接口原理图

21. SDCARD 接口

暴风四代开发板有一个 SDCARD 接口(见图 2.25),默认卡槽没有焊接,R14 有焊接,需要使用该功能的同学,可以自行焊接一个中号卡槽。

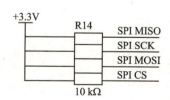

图 2.25　SDCARD 接口原理图

22. SDRAM 存储器

暴风四代开发板有一个 64 Mbit 的 SDRAM 芯片(见图 2.26)。

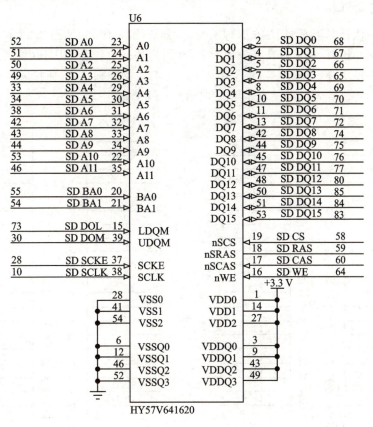

图 2.26　SDRAM 存储器原理图

该 SDRAM 使用 16 bit 接口，64 Mbit，可以作为数据和程序的存储器。Qsys/Nios II 系统也可以使用 SDRAM 作为程序存储器。

23. 配置芯片

暴风四代开发板有一个 4 Mbit 的 FPGA 配置 Flash 芯片——EPCS4 芯片(见图 2.27)。

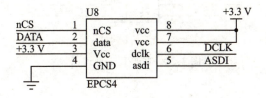

图 2.27　配置芯片原理图

EPCS4 可以存储 FPGA 的配置数据，保证 FPGA 掉电后程序不消失，使用 AS 下载方法可以把程序下载到 EPCS4 里面。

24. 扩展口

暴风四代开发板有两个 40 引脚扩展口，分别是 J3 和 J4（见图 2.28 和图 2.29），可以做通用的扩展，也可以插开发的配套外设；另外一个扩展口是 J8（见图 2.30），可以插摄像头模块、网络模块等；每个引脚在板上都有数字描述，方便扩展开发，在扩展开发的时候建议首先选用没有使用的 FPGA I/O 引脚。

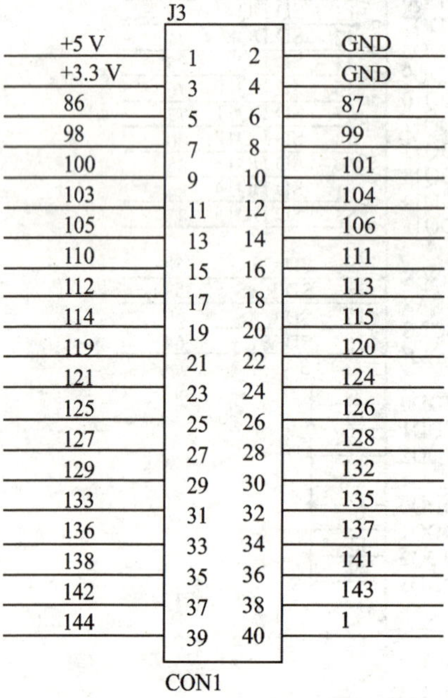

图 2.28 扩展口 1 原理图 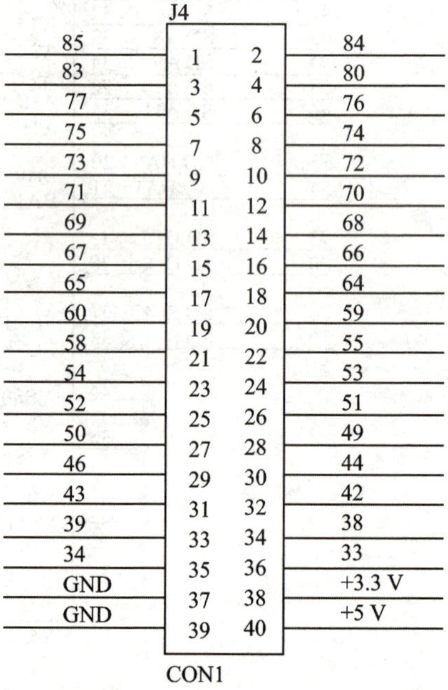   图 2.29 扩展口 2 原理图

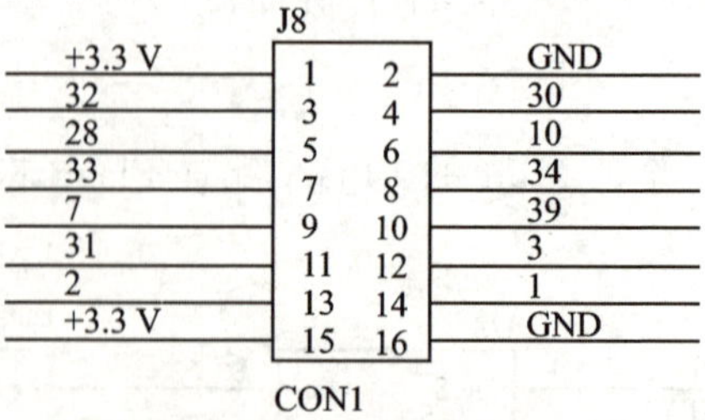

图 2.30 扩展口 3 原理图

第 3 章

设计实例

本章是书中最重要的一章,通过一系列由简单到复杂的实际例子,为您打开学习 FPGA 的大门,让您在 FPGA 的世界里遨游。切忌,眼高手低!先认真学习本书的例子,然后抛开所给的程序,自己编写相应的程序。看着自己的程序在板子上运行,难道不是一件非常有成就感的事情吗?

3.1 LED 流水灯实验

3.1.1 LED 简介

LED(Light Emitting Diode,发光二极管),是一种固态的半导体器件,它可以直接把电转化为光。此种元件在 1962 年出现,早期只能发出低光度的红光,被 HP 公司买下专利后当作指示灯。之后发展出其他单色光的版本,时至今日,能够发出的光已经遍及可见光、红外线及紫外线,光度亦提高到相当高的程度。LED 最开始用作指示灯。由于白光发光二极管的出现,LED 被普遍用作照明用途。

LED 的心脏是一个半导体的晶片,晶片的一端附在一个支架上,一端是负极,另一端连接电源的正极,整个晶片被环氧树脂封装起来。

LED 流水灯原理

让二极管形成流水,即先点亮一个 LED 灯,等待一小段时间后熄灭;再让第二个 LED 灯点亮,过一小段时间后再熄灭;以此类推,当最后一个 LED 灯点亮,过一小段时间后熄灭,然后再点亮第一个 LED 灯,如此循环。让 LED 发光比较简单,只需要 I/O 口电平发生变化即可。那么如何产生一个时间间隔呢?最常用的就是使用计数器延时来控制发光二极管。

3.1.2 实验任务

8个LED灯依次顺序点亮,产生流水效果,相邻LED灯发光时间间隔大约为0.5 s,人眼可以明显感知到这个间隔。

3.1.3 硬件设计

如图3.1所示,LED1~LED8连接到FPGA对应的I/O口上,通过控制I/O口的电平变化,实现二极管的导通,从而使发光二极管发光。以D1为例,当端口39为高电平(3.3 V)时,LED1发光;为低电平时,LED1熄灭。排阻R10和R13为限流电阻。排阻具有方向性,与色环电阻相比,具有整齐、所占空间少的优势。

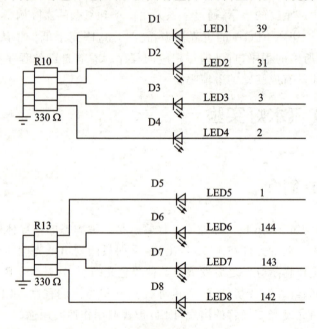

图3.1 LED流水灯实验原理图

3.1.4 程序设计

1. 设计思路

前面已经提到过,通过控制I/O口的高低电平可以实现LED发光或者熄灭,但是该怎样使用计数器实现延时呢?

实验要求相邻的LED时间间隔为0.5 s,FPGA开发板的晶振为50 MHz,所以有 0.5 s/20 ns＝25 000 000（2^{24}＜25 000 000＜2^{25}），需要 25 位计数位宽。当有效位计数达到最大值时，实现翻转（全 1 变全 0，往高一位进 1）。

本实验使用两个计数器。第一个计数器实现延迟间隔，计数器位宽为 25 bit，计数器命名为 counter；第二个计数器控制哪个 LED 灯亮，计数器位宽为 3 bit，计数器命名 count。

counter 会一直持续计数，计数到最大值后，会翻转为 0。当 counter 每计数至 0 的时候，count 加 1；当 count 计数到最大值后，也会翻转为 0。

2. 源代码

```
/******************Copyright (c)********************
**                      Adong   Studio
** ----------------------File Info----------------------
** File name:          LED
** Last modified Date: 2015-06-01
** Last Version:       1.1
** Descriptions:       LED
**
** ---------------------------------------------------
** Created by:         Adong
** Created date:       2015-06-01
** Version:            1.0
** Descriptions:       The original version
**
** ---------------------------------------------------
** Modified by:
** Modified date:
** Version:
** Descriptions:
**
** ---------------------------------------------------
*****************************************************/
module LED(
                         //input
        input            sys_clk,   //system clock
        input            sys_rst_n, //system reset,low is active
                         //output
        output reg[7:0]  LED
        );
//reg define
reg    [2:0]     count;
reg    [24:0]    counter;

//****************************************************
```

```verilog
// * *                        Main Program
// ***********************************************************/

//count for add counter,0.5 s/20 ns = 25 000 000,need 25 bit cnt
always @(posedge sys_clk or negedge sys_rst_n) begin
    if ( sys_rst_n == 1'b0 )
        counter <= 25'b0;
    else
        counter <= counter + 25'b1;
end

always @(posedge sys_clk or negedge sys_rst_n) begin
    if ( sys_rst_n == 1'b0 )
        count <= 3'b0;
    else if( counter == 25'd0 )
        count <= count + 3'b1;
    else ;
end

//ctr the LED pipiline display when count is equal to 0,1
always @(posedge sys_clk or negedge sys_rst_n) begin
    if ( sys_rst_n == 1'b0 )
        LED <= 8'b0;
    else begin
        case ( count )    //通过控制 I/O 口的高低电平实现发光二极管的亮灭
            0: LED = 8'b00000001;
            1: LED = 8'b00000010;
            2: LED = 8'b00000100;
            3: LED = 8'b00001000;
            4: LED = 8'b00010000;
            5: LED = 8'b00100000;
            6: LED = 8'b01000000;
            7: LED = 8'b10000000;
            default: LED = 8'b00000000;
        endcase
    end
end

endmodule
//end of rtl code
```

3.1.5 实验现象

将程序下载到开发板后,可观察到 D8~D1 从左至右轮流点亮,形成流水效果。LED 流水灯实验现象如图 3.2 所示。需要注意的是,当 D1 点亮后,会经过较长的时间 D8 才重新开始循环。

图 3.2 LED 流水灯实验现象

3.2 按键控制 LED 实验

3.2.1 按键控制 LED 简介

按键是最常用的输入设备,广泛用于单片机/FPGA/DSP 的输入控制,操作人员可以通过按键输入数据或者命令,实现简单的人机对话。FPGA 开发板使用的按键是一种常开型的开关,通常按键的两个触点不按下时处于断开状态,按下时处于闭合状态。

图 3.3 为按键实物图。

本实验使用按键控制发光二极管。发光二极管输入为高时发光,按键默认未按下时由于上拉电阻原因输出高电平,所以需要将按键输入取反后赋值给发光二极管输入端,即可控制发光二极管发光。按键未按下时 LED 灯处于熄灭状态,按键

图 3.3 按键实物图

按下时 LED 灯处于点亮状态。

3.2.2 实验任务

一个按键控制两个发光二极管,按下按键的可以使两个 LED 灯发光,未按下时两个 LED 灯不发光。

3.2.3 硬件设计

本实验需要按键和 LED 灯两个硬件外设,按键作为输入设备使用,4 个按键输入 4 个信号,LED 灯作为输出设备显示,1 个按键控制 2 个 LED 灯,4 个按键总共控制 8 个 LED 灯。R11 为上拉电阻,控制按键没有按下时,输入高电平。通用按键原理图如图 3.4 所示,LED 灯控制原理图如图 3.5 所示。

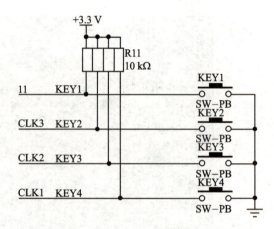

图 3.4 按键控制 LED 实验通用按键原理图

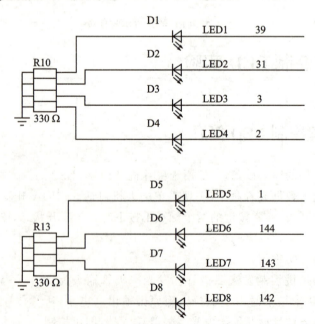

图 3.5 按键控制 LED 实验 LED 灯控制原理图

3.2.4 程序设计

1. 设计思路

输入为 4 个按键,输出为 8 个 LED,按键按下时按键输入为低电平,而 LED 需要驱动为高电平才能点亮,所以需要对按键输入进行取反,作为 LED 的输入。

2. 源代码

```
module KeyToLED(
//input
input           [3:0]       key,        //开发板上的 4 个按键控制 8 个 LED
//output
output wire     [7:0]       LED
//特别说明:什么时候用 wire 类型,什么时候用 reg 类型?
//判断原则:在使用 always 时,不管有没有时钟,都必须使用 reg 定义(一种是带时钟的
//always 块,一种是不带时钟的 always 块,不带时钟时使用阻塞赋值"=",带时钟时使
//用非阻塞赋值"<=";assign 使用 wire 类型定义
);

//reg define

//parameter define

/******************************************************
**                        Main Program
******************************************************/

// key 0 ctrl LED 0 & LED 4
// key 1 ctrl LED 1 & LED 5
// key 2 ctrl LED 2 & LED 6
// key 3 ctrl LED 3 & LED 7

// when key is pressed , key input low level, else key input high level
assign LED = ~{ key, key } ;                //key value display in beep
//即 assign LED = ~{key[3],key[2],key[1],key[0],key[3],key[2],key[1],key[0],};

endmodule
```

3.2.5 实验现象

如图 3.6 所示,将程序下载到开发板上,按下 KEY3,则相对应的 D3 和 D7 灯亮。

图 3.6 按键控制 LED 实验现象

3.3 七段数码管静态显示实验

3.3.1 数码管简介

数码管由 7 个条状和一个点状发光二极管制成,称为七段数码管。数码管是一种价格便宜、使用简单的经典显示器件。通过对其不同的引脚输入不同的电平,就可以让特定的数码段发光。

数码管在电器特别是家电领域应用十分广泛,如显示屏、空调、热水器、冰箱等。绝大多数热水器使用的都是数码管,也有的家电使用液晶屏或荧光屏。

数码管在电子开发领域用的也非常多,ARM/单片机/DSP/FPGA 等的开发都常用数码管做显示器件。

七段数码管实物图如图 3.7 所示。

1. 七段数码管原理

七段数码管由 7 个发光二极管组成,此外,还有一个圆点发光二极管(在图 3.8 中以 dp 表示),用于显示小数点。通过七段发光二极管亮暗的不同组合,可以显示多

第 3 章 设计实例

图 3.7 七段数码管实物图

种数字、字母以及其他符号。

2. LED 数码管分类

(1) 共阴极接法(也叫共阴数码管)

把发光二极管的阴极连在一起构成公共阴极。使用时公共阴极接地,这样阳极端输入高电平的段发光二极管就导通点亮,而输入低电平的段发光二极管则不点亮。

(2) 共阳极接法(也叫共阳数码管)

把发光二极管的阳极连在一起构成公共阳极。使用时公共阳极接+5 V。这样阴极端输入低电平的段发光二极管就导通点亮,而输入高电平的段发光二极管则不点亮。

数码管的逻辑图如图 3.8 所示,全部由发光二极管组合构成。

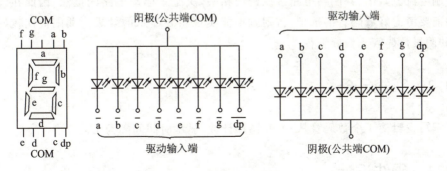

图 3.8 数码管逻辑图(1)

本实验使用的是共阳极数码管,例如:假如要显示数字 1,则 b、c 接低电平。

3. 数码管驱动方式

数码管有直流驱动和动态显示驱动两种驱动方式。直流驱动是指每个数码管的每一个段码都由一个单片机或者 FPGA 的 I/O 端口进行驱动,或者使用如 BCD 码二-十进制译码器译码进行驱动。优点是编程简单,显示亮度高,缺点是占用 I/O 端口多。动态显示驱动是将所有数码管通过分时轮流控制各个数码管的 COM 端,使各个数码管轮流受控显示。将所有数码管的 8 个显示笔画"a,b,c,d,e,f,g,dp"的同名端连在一起;另外,为每个数码管的公共极 COM 端增加位选通控制电路。位选通由各自独立的 I/O 线控制,当单片机或者 FPGA 的 I/O 口输出字形码时,所有数码管都接收到相同的字形码,但究竟是哪个数码管会显示出字形,取决于单片机或者 FPGA 对位选通 COM 端电路的控制。所以我们只要将需要显示的数码管的选通控制打开,该位就会显示出字形,没有选通的数码管则不会亮。本实验利用 FPGA 的 I/O 口及三极管来驱动数码管,为静态直流驱动方式。

4. 显示举例

如果用图 3.9 的数码管显示数字 8,那么图 3.9 中

e g f dp c d b a

应该按照

1 1 1 0 1 1 1 1

显示。

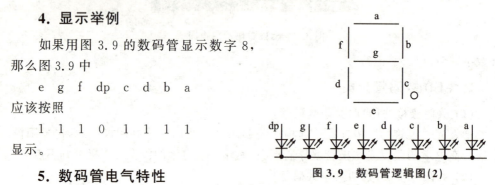

图 3.9 数码管逻辑图(2)

5. 数码管电气特性

LED 数码管中各段发光二极管的伏安特性与普通二极管类似,只是正向压降与正向电阻较大。在一定范围内,其正向电流与亮度成正比。由于常规数码管的起辉电流只有 1~2 mA,最大限电流也只有 10~30 mA,所以我们在实验中使用了排阻来限制电流。共阳、共阴与电路接线密切相关,决定了驱动电路的接法,因此在设计电路前要考虑好数码管的选型,否则就不能实现想要显示的效果。共阳极比较好驱动,共阴极数码管比较好编程。

3.3.2 实验任务

编写逻辑使 4 个数码管从 0~9 循环计数,显示时间间隔为 1 s。

3.3.3 硬件设计

数码管硬件由 4 个 PNP 三极管、3 个限流电阻和 1 个八段 4 位的数码管组成;

R17、R19 和 R18 为限流电阻，R17 和 R19 的电阻值为为 330 Ω，是数码管实测亮度比较合适的阻值。

各位数码管的共阳极由 FPGA 控制 Q9～Q12 来实现 4 位数码管的位输出。本实验数码管是共阳极。

静态显示的数码管原理图如图 3.10 所示。对于数码管静态显示，C1/C2/C3/C3 位选信号统一进行控制，根据需要将显示的数字送入相应的段码（A－B－C－D－E－F－G－H），即可让 4 位数码管显示出相同的数字。

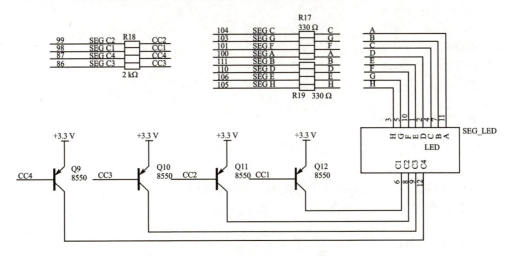

图 3.10　七段数码管静态显示实验原理图

3.3.4　程序设计

1. 设计思路

逻辑主要由两部分构成：一个是对时间的控制，通过控制数码管跳变的时间，实现四位数码管从 0 到 9 的循环，该部分可由计数器实现；另外一个是对数码管译码的控制，使数码管循环显示 0～9 这 10 个数字。

2. 源代码

```
module SegLED (
//input
input    sys_clk,
input    sys_rst_n,

//output
output wire    seg_c1,            //输入低电平使三极管导通
```

```verilog
    output wire    seg_c2,
    output wire    seg_c3,
    output wire    seg_c4,

    output reg    seg_a,
    output reg    seg_b,
    output reg    seg_c,
    output reg    seg_e,
    output reg    seg_d,
    output reg    seg_f,
    output reg    seg_g,
    output reg    seg_h
);

//parameter define
parameter SIZE = 8;

//reg define
reg [3:0]         counter;   //控制逻辑产生 0～9 数字
reg [SIZE-1:0]    disp_data;

reg    disp_clk;

reg [25:0]        clk_cnt;   //控制时间,$2^{26}$ = 67 108 864＞50 000 000

reg    segled_a;    //数码管段位定义
reg    segled_b;
reg    segled_c;
reg    segled_e;
reg    segled_d;
reg    segled_f;
reg    segled_g;
reg    segled_h;

//wire define

/******************************************************
**                    Main Program
******************************************************/

//使用计数器控制数码管延时,复位或计数到 1 s 时,重新开始计数。1 s = 50 000 000 × 20 ns
always @(posedge sys_clk or negedge sys_rst_n) begin
```

```verilog
    if (sys_rst_n == 1'b0)
        clk_cnt <= 26'b0;
    else if ( clk_cnt == 26'd50000000 )
        clk_cnt <= 26'b0;
    else
        clk_cnt <= clk_cnt + 26'b1;
end

//利用 clk_cnt 控制 counter 产生,实现 0~9 循环,用来控制数码管显示。RESET 清零
always @(posedge sys_clk or negedge sys_rst_n) begin
    if (sys_rst_n == 1'b0)
        counter <= 4'd0;
    else if ( clk_cnt == 26'd50000000 && counter == 4'd9 )
        counter <= 4'b0;
    else if ( clk_cnt == 26'd50000000 )
        counter <= counter + 4'b1;
    else ;
end

//对位段进行译码,显示 0~9
// control    SEGLED disp 0 - 9
always @( * ) begin
    case (counter)
            9:
                begin
                    segled_a = 1;
                    segled_b = 1;
                    segled_c = 1;
                    segled_e = 0;
                    segled_d = 0;
                    segled_f = 1;
                    segled_g = 1;
                    segled_h = 0;
                end
            8:
                begin
                    segled_a = 1;
                    segled_b = 1;
                    segled_c = 1;
                    segled_e = 1;
                    segled_d = 1;
                    segled_f = 1;
```

```
                segled_g = 1 ;
                segled_h = 1 ;
            end
    7:
        begin
            segled_a = 1 ;
            segled_b = 1 ;
            segled_c = 1 ;
            segled_e = 0 ;
            segled_d = 0 ;
            segled_f = 0 ;
            segled_g = 0 ;
            segled_h = 0 ;
        end
    6:
        begin
            segled_a = 1 ;
            segled_b = 0 ;
            segled_c = 1 ;
            segled_e = 1 ;
            segled_d = 1 ;
            segled_f = 1 ;
            segled_g = 1 ;
            segled_h = 0 ;
        end
    5:
        begin
            segled_a = 1 ;
            segled_b = 0 ;
            segled_c = 1 ;
            segled_e = 0 ;
            segled_d = 1 ;
            segled_f = 1 ;
            segled_g = 1 ;
            segled_h = 0 ;
        end
    4:
        begin
            segled_a = 0 ;
            segled_b = 1 ;
            segled_c = 1 ;
            segled_e = 0 ;
```

```
                segled_d = 0 ;
                segled_f = 1 ;
                segled_g = 1 ;
                segled_h = 0 ;
            end
3：
            begin
                segled_a = 1 ;
                segled_b = 1 ;
                segled_c = 1 ;
                segled_e = 0 ;
                segled_d = 1 ;
                segled_f = 0 ;
                segled_g = 1 ;
                segled_h = 0 ;
            end
2：
            begin
                segled_a = 1 ;
                segled_b = 1 ;
                segled_c = 0 ;
                segled_e = 1 ;
                segled_d = 1 ;
                segled_f = 0 ;
                segled_g = 1 ;
                segled_h = 0 ;
            end
1：
            begin
                segled_a = 0 ;
                segled_b = 1 ;
                segled_c = 1 ;
                segled_e = 0 ;
                segled_d = 0 ;
                segled_f = 0 ;
                segled_g = 0 ;
                segled_h = 0 ;
            end
0：
            begin
                segled_a = 1 ;
                segled_b = 1 ;
```

```
                    segled_c = 1 ;
                    segled_e = 1 ;
                    segled_d = 1 ;
                    segled_f = 1 ;
                    segled_g = 0 ;
                    segled_h = 1 ;
            end
        default:
            begin
                    segled_a = 0 ;
                    segled_b = 0 ;
                    segled_c = 0 ;
                    segled_e = 0 ;
                    segled_d = 0 ;
                    segled_f = 0 ;
                    segled_g = 0 ;
                    segled_h = 0 ;
            end
        endcase
end

//组合逻辑进行取反,数码管为共阳极,对译码段位取反,低电平有效
always @( * ) begin
    seg_a = ～segled_a ;
    seg_b = ～segled_b ;
    seg_c = ～segled_c ;
    seg_e = ～segled_e ;
    seg_d = ～segled_d ;
    seg_f = ～segled_f ;
    seg_g = ～segled_g ;
    seg_h = ～segled_h ;
end

//4个片选全部选通,4个片选输出为0,控制三极管,驱动数码管的4个片选全部为1
assign seg_c1 = 1'b0;
assign seg_c2 = 1'b0;
assign seg_c3 = 1'b0;
assign seg_c4 = 1'b0;

endmodule
//end of RTL code
```

3.3.5 实验现象

程序下载到开发板后,可以看到数码管从 0~9 循环显示,如图 3.11 所示。

图 3.11 七段数码管静态显示实验现象

3.4 七段数码管动态扫描实验

3.4.1 动态扫描简介

动态扫描就是轮流向各位数码管送出字形码和响应的位选,只要扫描显示速度足够快,利用发光管的余辉和人眼视觉暂留作用,使人感觉好像各位数码管都在显示。

动态显示的特点是将所有位数码管的段选线并联在一起,由位选线控制是哪一位数码管有效。这样一来,就没有必要为每一位数码管配一个锁存器,从而大大地简化了硬件电路。选亮数码管采用动态扫描显示。动态显示的亮度比静态显示要差一些,所以在选择限流电阻时应略小于静态显示电路中的限流电阻。本实验中的限流电阻和静态的一样,影响不是很大。

数码管不同位显示的时间间隔可以通过调整延时程序的延时长短来完成。数码管显示的时间间隔还能够确定数码管显示时的亮度。若显示的时间间隔长,则显示时数码管的亮度将亮些;若显示的时间间隔短,则显示时数码管的亮度将暗些;若显示的时间间隔过长,则数码管显示时将产生闪烁现象。所以,在调整显示的时间间隔时,既要考虑显示时数码管的亮度,又要考虑数码管显示时不闪烁的现象。一般控制在 1 ms 左右最佳。本实验设计开发板的刷新时间为 1 ms。

3.4.2 实验任务

数码管的 4 位动态显示每个不同的值。

3.4.3 硬件设计

七段数码管动态扫描实验原理图如图 3.12 所示。对于数码管动态扫描,需要动态选通 C1/C2/C3/C3 位选信号,每个位选通时根据该位需要显示的数字送入相应的段码(A—B—C—D—E—F—G—H),依次轮询扫描显示,即可显示出动态效果。

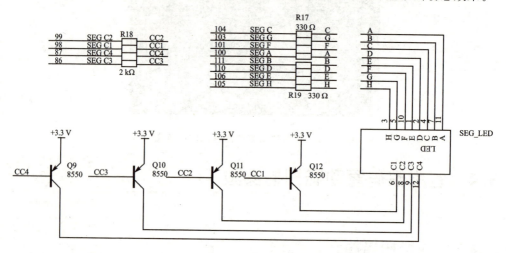

图 3.12 七段数码管动态扫描实验原理图

3.4.4 程序设计

1. 设计思路

数码管动态扫描显示需要几个计数器来控制:
- scan_cnt 扫描计数器,用来动态选通 4 个位选信号,形成动态显示效果。
- clk_cnt 计数器,控制每位数码管的显示延迟。

- counter 计数器,产生 0~9 个数字,控制数码管的显示数字。

2. 源代码

```
module SegLed_DinamDisp(
//input
input               sys_clk,
input               sys_rst_n,

//output
output wire         seg_c1,
output wire         seg_c2,
output wire         seg_c3,
output wire         seg_c4,

output reg          seg_a,
output reg          seg_b,
output reg          seg_c,
output reg          seg_e,
output reg          seg_d,
output reg          seg_f,
output reg          seg_g,
output reg          seg_h
);

//parameter define
parameter SIZE = 8;

//reg define
reg     [3:0]       counter;
reg     [3:0]       disp_data;

reg                 disp_clk;

reg     [25:0]      clk_cnt;

reg     [15:0]      scan_cnt;
reg     [3:0]       segled_bit_sel;

reg                 segled_a;
reg                 segled_b;
reg                 segled_c;
```

```verilog
    reg              segled_e;
    reg              segled_d;
    reg              segled_f;
    reg              segled_g;
    reg              segled_h;

//wire define

/***************************************************
**                Main Program
***************************************************/
//产生一个 1 ms 左右的计数,作为动态选通 4 个位选信号
always @(posedge sys_clk or negedge sys_rst_n) begin
    if (sys_rst_n == 1'b0)
        scan_cnt <= 16'b0;
    else
        scan_cnt <= scan_cnt + 16'b1;      //20 ns × 2^16 = 1 310 720 ns≈1.3 ms
end

//产生数码管位选,使用 scan_cnt 高 2 bit 进行控制
always @(posedge sys_clk or negedge sys_rst_n) begin
    if (sys_rst_n == 1'b0)                 //复位时从数码管低位开始扫描
        segled_bit_sel <= 4'b0001;
    else if ( scan_cnt[15:14] == 2'b00 )
//扫描计数 0000 0000 0000 0000~1111 1111 1111 1100,即每个扫描周期从 0 至 0.325 ms 时,
//点亮第一个数码管
        segled_bit_sel <= 4'b0001;
    else if ( scan_cnt[15:14] == 2'b01 )
//0.325~0.65 ms 时,点亮第二个数码管
        segled_bit_sel <= 4'b0010;
    else if ( scan_cnt[15:14] == 2'b10 )
//0.65~0.975 ms 时,点亮第三个数码管
        segled_bit_sel <= 4'b0100;
    else if ( scan_cnt[15:14] == 2'b11 )
//0.975~1.3 ms 时,点亮第四个数码管
        segled_bit_sel <= 4'b1000;
    else ;
end

//使用计数器控制数码管延时,复位或计数到 1 s 时,重新开始计数。1 s = 50 000 000 × 20 ns
always @(posedge sys_clk or negedge sys_rst_n) begin
    if (sys_rst_n == 1'b0)
```

```
        clk_cnt <= 26'b0;
    else if ( clk_cnt == 26'd50000000 )
        clk_cnt <= 26'b0;
    else
        clk_cnt <= clk_cnt + 26'b1;
end

//利用 clk_cnt 控制 counter 产生,实现 0~9 循环,用来控制数码管显示。RESET 清零
always @(posedge sys_clk or negedge sys_rst_n) begin
    if (sys_rst_n == 1'b0)
        counter <= 4'd0;
    else if ( clk_cnt == 26'd50000000 && counter == 4'd9 )
        counter <= 4'b0;
    else if ( clk_cnt == 26'd50000000 )
        counter <= counter + 4'b1;
    else ;
end

//让数码管 4 个位同时输出不同的数字
always @( * ) begin
    if ( segled_bit_sel == 4'b0001 )
        disp_data = counter + 4'd1 ;       //低 1 位接收到高电平时加 1
    else if ( segled_bit_sel == 4'b0010 )
        disp_data = counter + 4'd2 ;       //低 2 位数码管接收到高电平时加 2
    else if (segled_bit_sel == 4'b0100 )
        disp_data = counter + 4'd3 ;       //低 3 位数码管接收到高电平时加 3
    else
        disp_data = counter + 4'd4 ;       //低 4 位数码管接收到高电平时加 4
end

//对位段进行译码,显示 0~9
// control    SEGLED disp 0 - 9
always @( * ) begin
    case (counter)
        9:
                begin
                    segled_a = 1 ;
                    segled_b = 1 ;
                    segled_c = 1 ;
                    segled_e = 0 ;
                    segled_d = 0 ;
                    segled_f = 1 ;
                    segled_g = 1 ;
                    segled_h = 0 ;
                end
        8:
```

```
            begin
                segled_a = 1 ;
                segled_b = 1 ;
                segled_c = 1 ;
                segled_e = 1 ;
                segled_d = 1 ;
                segled_f = 1 ;
                segled_g = 1 ;
                segled_h = 1 ;
            end
        7:
            begin
                segled_a = 1 ;
                segled_b = 1 ;
                segled_c = 1 ;
                segled_e = 0 ;
                segled_d = 0 ;
                segled_f = 0 ;
                segled_g = 0 ;
                segled_h = 0 ;
            end
        6:
            begin
                segled_a = 1 ;
                segled_b = 0 ;
                segled_c = 1 ;
                segled_e = 1 ;
                segled_d = 1 ;
                segled_f = 1 ;
                segled_g = 1 ;
                segled_h = 0 ;
            end
        5:
            begin
                segled_a = 1 ;
                segled_b = 0 ;
                segled_c = 1 ;
                segled_e = 0 ;
                segled_d = 1 ;
                segled_f = 1 ;
                segled_g = 1 ;
                segled_h = 0 ;
            end
        4:
            begin
                segled_a = 0 ;
                segled_b = 1 ;
```

```
                segled_c = 1 ;
                segled_e = 0 ;
                segled_d = 0 ;
                segled_f = 1 ;
                segled_g = 1 ;
                segled_h = 0 ;
            end
    3：
            begin
                segled_a = 1 ;
                segled_b = 1 ;
                segled_c = 1 ;
                segled_e = 0 ;
                segled_d = 1 ;
                segled_f = 0 ;
                segled_g = 1 ;
                segled_h = 0 ;
            end
    2：
            begin
                segled_a = 1 ;
                segled_b = 1;
                segled_c = 0 ;
                segled_e = 1 ;
                segled_d = 1 ;
                segled_f = 0 ;
                segled_g = 1 ;
                segled_h = 0 ;
            end
    1：
            begin
                segled_a = 0 ;
                segled_b = 1 ;
                segled_c = 1 ;
                segled_e = 0 ;
                segled_d = 0 ;
                segled_f = 0 ;
                segled_g = 0 ;
                segled_h = 0 ;
            end
    0：
            begin
                segled_a = 1 ;
                segled_b = 1 ;
                segled_c = 1 ;
                segled_e = 1 ;
                segled_d = 1 ;
```

```
                    segled_f = 1;
                    segled_g = 0;
                    segled_h = 1;
                end
            default:
                begin
                    segled_a = 0;
                    segled_b = 0;
                    segled_c = 0;
                    segled_e = 0;
                    segled_d = 0;
                    segled_f = 0;
                    segled_g = 0;
                    segled_h = 0;
                end
        endcase
end

//组合逻辑进行取反,数码管为共阳极,对译码段位取反,低电平有效
always @(*) begin
    seg_a = ~segled_a;
    seg_b = ~segled_b;
    seg_c = ~segled_c;
    seg_e = ~segled_e;
    seg_d = ~segled_d;
    seg_f = ~segled_f;
    seg_g = ~segled_g;
    seg_h = ~segled_h;
end

//4个片选通过 segled_bit_sel 进行选通,输出控制三极管,驱动数码管的4个片选
assign seg_c1 = ~( segled_bit_sel == 4'b0001 );
assign seg_c2 = ~( segled_bit_sel == 4'b0010 );
assign seg_c3 = ~( segled_bit_sel == 4'b0100 );
assign seg_c4 = ~( segled_bit_sel == 4'b1000 );

endmodule
//end of RTL code
```

3.4.5 实验现象

将程序下载到发板以后,可以看到数码管动态显示结果,4位数码管分别显示不同的数字,如图3.13所示。

图 3.13 七段数码管动态扫描实验现象

3.5 串口发送实验

3.5.1 串口简介

UART 是一种通用串行数据总线,用于异步通信。该总线双向通信,可以实现全双工传输和接收。一般较老的台式计算机机箱会配有 UART 串口,笔记本式计算机和较新的台式计算机都取消了 UART 串口。

UART 串口是电子设计用得非常多的一种接口,UART 串口通信实现原理比较简单,基本所有的 MCU/ARM/DSP 等芯片都配置有串口,经常被用来作为调试接口,还可以作为 PC 和 CPU 通信的低速接口。

计算机里面的台式机 DB9 串口如图 3.14 所示。

由于很多计算机都没有 UART 串口,所以串口通信可以使用 USB 转串口线(一端是 USB 口,一端是 DB9 串口,方便没有 DB9 串口的计算机使用 USB 接口实现 DB9 串口功能)实现,USB 转串口接口和计算机的串口使用起来是一样的,接口如图 3.15 所示。

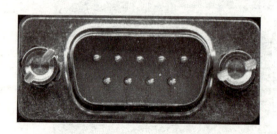
图3.14 台式机串口实物图

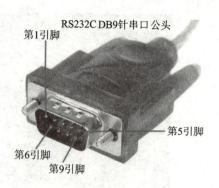
图3.15 USB串口实物图

DB9的串口定义如表3.1所列,一般只使用引脚2(RXD)和引脚3(TXD)。

表3.1 DB9串口接口定义

引 脚	简 写	功能说明
1	CD	载波侦测(Carrier Detect)
2	RXD	接收数据(Receive)
3	TXD	发送数据(Transmit)
4	DTR	数据终端准(Data Terminal Ready)
5	GND	地线(Ground)

1. 波特率

串口有波特率的概念,波特率是表示串口发送速率的一个单位,一般串口的波特率为3 200、6 400、9 600、15 200等,波特率9 600表示每秒传输9 600bit(位)数据,即9 600 bps。

2. 名词解释

UART(Universal Asynchronous Receiver and Transmit)是通用异步收发器,是通信上的传输模式。

RS232是一种异步串行通信协议标准,最初是为远程通信连接数据终端设备DTE(Data Terminal Equipment)与数据通信设备DCE(Data Communication Equipment)而制定的。

RS232标准规定了连接电缆和机械、电气特性、信号功能及传送过程,但是RS232并未定义连接器的物理特性,因此存在DB25、DB15和DB9各种类型的连接器,其引脚的定义也是各不相同的。

DB9是具体的物理电气接口器(connector),一般使用的RS232电气连接器都是DB9的。

3.5.2 实验任务

实现 9 600 波特率的串口发送,按下按键发送一次数据(8 bit,一个字节)。上位机(PC)串口软件可以接收到发送的数据。

3.5.3 硬件设计

本实验主要使用 DB9 串口的 2、3 引脚,两个串口连接时,接收数据引脚与发送数据引脚相连,彼此交叉,信号对应接地即可。

RS232C 在电气特性上表现为传送的数字量采用负逻辑,且与地对称,所以和单片机或者 FPGA 开发板连接时常常需要加入电平转换芯片 MAX3232C。

MAX3232C 为电平转换芯片,将 TTL 电平转换为 RS232 电平。MAX3232 等收发器是采用专有的低压差发送器输出级,利用双电荷泵在 3.0~5.5 V 电源供电。能够实现真正的 RS232 性能,MAX3232C 供电电压为 3.3 V。

串口发送实验原理图如图 3.16 所示。

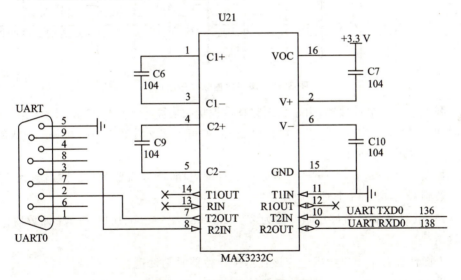

图 3.16 串口发送实验原理图

3.5.4 程序设计

1. 设计思路

以波特率 9 600 为例子进行说明,波特率 9 600 接收 1 bit 的时间为 1 s/9 600 =

104 μs(微秒),即每隔 104 μs 发送一位数据。

图 3.17 是串口一数据帧(发送 8 bit 有效数据)时序图,串口都是一个字节一个字节发送数据的,每个字节都需要占用一次传输,包括起始位(1 bit)、数据位(8 bit)、校验位(校验数据正确性,1 bit)、停止位(1 bit)。

图 3.17 串口数据帧时序图(1)

下面分析把数据按照开发板的系统时钟 50 MHz 通过串口发送出去。

104 μs = 104 000 ns,50 MHz 时钟的一个周期时间为 20 ns(1/50 MHz=20 ns),一个时钟周期远小于 104 000 ns,所以可以用 50 MHz 时钟产生的计数器来计数,产生串口的发送时序,然后根据计数器特定的值来发送数据。

图 3.18 中的竖线为各个位可以发送数据的起始,总共有 11 个数据位(1 个起始位+8 个数据位+1 个校验位+1 个停止位)要发送。

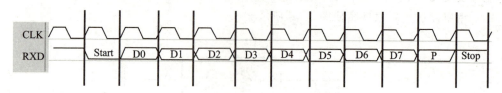

图 3.18 串口数据帧时序图(2)

我们新建一个计数器,由于一次完整数据接收需要有 1 144 000 ns(104 000×11)的时间,所以计数器计到大于 1 144 000 ns 时即可。那么计数 1 144 000 ns 需要多少个 50 MHz 的时钟周期呢?

需要 1 144 000 ns/20 ns = 57 200 个时钟周期,所以需要 16 位计数器,16 位最大计数到 2^{16}=65 536 个时钟周期。

15 bit 计数器计数够用吗?因为 2^{15}=32 768 < 57 200,所以 15 bit 计数器位数不够用。

下面计算各个串口数据位应该在哪个计数器时刻发送。

Startbit(0)	占据计数器的 0~5 200(=104 000 ns/20 ns)
D0	占据计数器的 5 200~10 400(=208 000 ns/20 ns)
D1	占据计数器的 10 400~15 600(=312 000 ns/20 ns)
D2	占据计数器的 15 600~20 800(=416 000 ns/20 ns)
D3	占据计数器的 20 800~26 000(=520 000 ns/20 ns)
D4	占据计数器的 26 000~31 200(=624 000 ns/20 ns)
D5	占据计数器的 31 200~36 400(=728 000 ns/20 ns)

D6　　　　　　　　占据计数器的 36 400～41 600(=832 000 ns/20 ns)
D7　　　　　　　　占据计数器的 41 600～46 800(=936 000 ns/20 ns)
校验位　　　　　　占据计数器的 46 800～52 000(=10 040 000 ns/20 ns)
Endbit(1)　　　　 占据计数器的 52 000～57 200(=1 144 000 ns/20 ns)

根据上面的计数间隔发送对应的串口数据就可以了,就是 9 600 波特率的数据发送时序。

2. 按键去抖

本实验我们通过按键发送数据,按键所用开关为机械弹性开关,当机械触点断开、闭合时,由于机械触点的弹性作用,所以一个按键开关在闭合时不会马上稳定地接通,在断开时也不会立刻断开。因而在闭合及断开的瞬间均伴随一连串的抖动,会使输入产生毛刺,导致发送值不稳定。为了不产生这种现象而作的措施就是按键消抖,如图 3.19 所示。

抖动时间的长短由按键的机械特性决定,一般为 5～10 ms。这是一个很重要的时间参数,在很多场合都要用到。按键稳定闭合时间的长短是由操作人员的按键动作决定的,一般为零点几秒至数秒。键抖动会引起一次按键被误读多次。为确保 CPU/FPGA 对键的一次闭合仅做一次处理,必须去除键抖动。在键闭合稳定时读取键的状态,并且必须判别到键释放稳定后再做处理。

通常按键去抖时有软件去抖和硬件去抖两种方式,本实验使用软件去抖。即按键处于稳定状态时,才开始发送数据。因为发送数据只需计数器计数到 57 200,这个时间还是相当短的,可以做到数据在去抖后马上发送完,不会受到按键后沿的影响,所以不需要考虑后沿抖动。

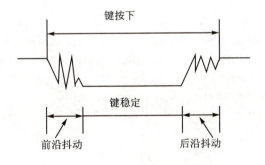

图 3.19　按键去抖示意图

3. 源代码

```
module UartSend(
    input    wire           sys_clk,      //system clock
    input    wire           sys_rst_n,    //system reset, low is active
    input              [3:0] key_data,    //KEY data in 4 bit
```

```verilog
    output   wire    [7:0]      LED,          //led output
    output   reg                uart_txd
                 )

//parameter define
parameter WIDTH = 8;
parameter SIZE = 16;
parameter DELAY_CNT = 10000000;

//reg define
reg  [WIDTH-1:0]    buff;
reg  [WIDTH-1:0]    data_out;

reg                 uart_txd;
reg  [WIDTH-1:0]    txd;            //temp txd signal

reg  [SIZE-1:0]     counter;
reg  [23:0]         delay_count;

//wire define
wire                sys_clk;
wire                sys_rst_n;

/******************************************************************
**                           Main Code
******************************************************************/
//将按键值赋给buff,由于buff是8位,key_dada是4位,这里相当于将key_data拼接成
//8位,即{key_data[3],key_data[2],key_data[1],key_data[0],key_data[3],key_data[2],
//key_data[1],key_data[0]}
always @(posedge sys_clk or negedge sys_rst_n) begin
        if (sys_rst_n == 1'b0)              //初始化buff,复位时全部清零
            buff <= 8'b0;
        else
            buff <= { key_data , key_data };
end

//将buff赋值给LED,使数值在LED输出显示
assign LED = buff;

//使用软件去抖
always @(posedge sys_clk or negedge sys_rst_n) begin
```

```verilog
        if ( sys_rst_n == 1'b0)
            delay_count <= 24'b0;
        else if (key_data ! = 4'hf && delay_count <= DELAY_CNT )
//去抖时间 DELAT_CNT = 10 000 000 × 20 ns = 0.2 s,需要 24 位计数器。任意按键按下且
//在 0.2 s 之内,开始计数
            delay_count <= delay_count + 24'b1;
        else
            delay_count <= 24'b0;
end

always @(posedge sys_clk or negedge sys_rst_n ) begin
        if ( sys_rst_n == 1'b0 )
            counter <= 16'b0;
        else if (delay_count == DELAY_CNT)
//counter 在 delay_count 计数到 DELAY_CNT 时开始发送数据
            counter <= 16'b0;
        else if (counter <= 57200)                    //完成一次发送数据
            counter <= counter + 16'b1;
        else ;
end

always @( * ) begin
    if ((counter > 0)&&(counter<= 5200 ))            //发送 start
      txd = 1'b0;
    else if ((counter > 5200) &&(counter<= 10400))   //发送 D0
      txd = buff[0];
    else if ((counter > 10400) && (counter<= 15600)) //发送 D1
      txd = buff[1];
    else if ((counter > 15600) && (counter<= 20800)) //发送 D2
      txd = buff[2];
    else if ((counter > 20800) && (counter<= 26000)) //发送 D3
      txd = buff[3];
    else if ((counter > 26000) && (counter<= 31200)) //发送 D4
      txd = buff[4];
    else if ((counter > 31200) && (counter<= 36400)) //发送 D5
      txd = buff[5];
    else if ((counter > 36400) && (counter<= 41600)) //发送 D6
      txd = buff[6];
    else if ((counter > 41600) && (counter<= 46800)) //发送 D7
      txd = buff[7];
    else if ((counter > 46800) && (counter<= 52000)) //校验位
      txd = 1'b1;
```

```verilog
        else if ((counter > 52000) && (counter <= 57200))    //停止位
            txd = 1'b1;
        else
            txd = 1'b1;
    end

always @(posedge sys_clk or negedge sys_rst_n) begin
        if (sys_rst_n == 1'b0)
            uart_txd <= 1'b1;
        else
            uart_txd <= txd;              //寄存器输出,打拍处理,防止时序毛刺产生

    end

endmodule
//end of RTL code
```

3.5.5 实验现象

打开图 3.20 的串口软件(光盘中会配送,仅作学习使用),将程序下载到开发板,连接好普通串口线或者 USB 串口线后(如果是 USB 串口,那么图 3.20 所示的串口软件的串口号可能为 COM3 或者 COM4),并勾选 HEX 显示(十六进制显示)。按下开发板的 key1 至 key4 后会显示响应的值。

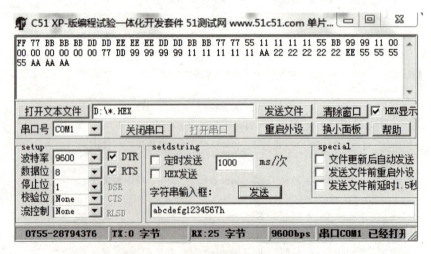

图 3.20 串口发送实现软件界面

3.6 串口接收实验

3.6.1 串口接收简介

串口接收是指将上位机通过串口发送的数据接收下来,存储在 FPGA 逻辑内部,然后显示到 LED 灯上面。因为本书采用的开发板的 LED 灯只有 8 个,所以该实验只能显示上位机的一个字节数据(一个字节数据有 8 bit 内容)。

3.6.2 实验任务

实现波特率 9 600 的串口接收,上位机串口软件发送的数据,可以被接收到,并通过 LED 显示出来。

3.6.3 硬件设计

串口接收是将上位机通过串口发送的数据接收下来,UART 串口连到上位机上,然后使用串口软件发送数据,串口模块就能接收到对应的数据,通过 MAX3232 电平转换芯片后,FPGA 即可收到对应的数据。串口接收实验原理图如图 3.21 所示。

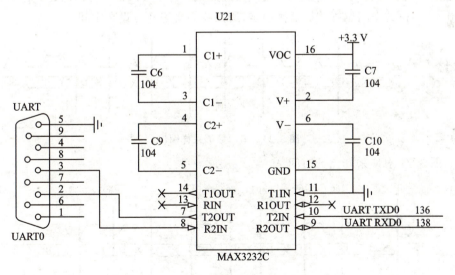

图 3.21 串口接收实验原理图

3.6.4 程序设计

1. 设计思路

以波特率 9 600 为例子进行说明，波特率 9 600 接收 1 bit 的时间为 1 s/9 600＝104 μs，即每隔 104 μs 接收一位数据。

图 3.22 是串口一数据帧（接收 8 bit 有效数据）时序图，串口都是一个字节一个字节接收数据的，每个字节都需要占用一次传输，包括起始位（1 bit）、数据位（8 bit）、校验位（校验数据的正确性，1 bit）、停止位（1 bit）。

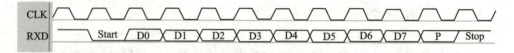

图 3.22 串口一数据帧时序图

下面分析把数据按照开发板的系统时钟频率 50 MHz 通过串口发送出去。

104 μs ＝ 104 000 ns，50 MHz 时钟频率的一个周期为 20 ns（1/50 MHz＝20 ns），一个时钟周期远小于 104 000 ns，所以可以用 50 MHz 时钟频率产生的计数器来计数，产生串口的接收时序，然后根据计数器特定的值来采样数据。

串口数据帧起始位可以用一个下降沿检测电路来检测下降沿。

采样数据有个基本要求，就是要在数据稳定时采样，这样可以最好地满足寄存器的建立时间和保持时间，得到一个稳定的值，从时序图 3.23 上可以看出，在数据位（D0～D7）位的中间位置数据最为稳定，所以我们将采样点设在中间。

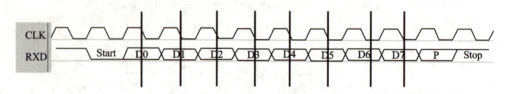

图 3.23 串口数据采样图(1)

图 3.24 对应采样点的时间

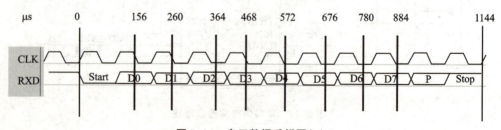

图 3.24 串口数据采样图(2)

新建一个计数器,50 MHz 为时钟,那么计数器变化一次时间为 20 ns。由于一次完整数据接收需要有 1 144 000 ns 的时间,所以计数器必须计到 1 144 000 ns。

一个串口数据(包括起始位(1 bit)、数据位(8 bit)、校验位(校验数据正确性,1 bit)、停止位(1 bit),共 11 bit)发送周期是 1 144 000 ns,那么每个串口发送位周期需要 1 144 000/11=104 000,图 3.24 从 0 开始到 D0 的采样时刻有 1.5 个时钟周期,所以有 1.5×104 000=156 000 ns(图 3.24 中的数字后省去了 000,其他采样位依次类推)。

各个计数位的采样时刻计算如下:
D0 对应计数器的 7 800 =(156 000 ns/20 ns)
D1 对应计数器的 13 000 =(260 000 ns/20 ns)
D2 对应计数器的 18 200 =(364 000 ns/20 ns)
D3 对应计数器的 23 400 =(468 000 ns/20 ns)
D4 对应计数器的 28 600 =(572 000 ns/20 ns)
D5 对应计数器的 33 800 =(676 000 ns/20 ns)
D6 对应计数器的 39 000 =(780 000 ns/20 ns)
D7 对应计数器的 44 200 =(884 000 ns/20 ns)

2. 源代码

```
module UartRecv(
                //input
                sys_clk,
                sys_rst_n,
                uart_rxd,
                
                //output
                LED
                );


//parameter define
parameter WIDTH = 8;
parameter SIZE = 16;

//input ports

    input           sys_clk;         //system clock
    input           sys_rst_n;       //system reset, low is active
    input           uart_rxd;        //uart rxd input
```

```verilog
//output ports
output reg [WIDTH-1:0]      LED;

//reg define
reg   [WIDTH-1:0]       buff;
reg   [WIDTH-1:0]       data_out;

reg                     uart_rxd_dly1;  //uart rxd input
reg                     uart_rxd_dly2;  //uart rxd input
reg                     uart_rxd_dly3;  //uart rxd input
reg                     uart_rxd_dly4;  //uart rxd input

reg   rxd_negdge_sig_dly1 ;

reg   [SIZE-1:0]        counter;

//wire define
wire                    rxd_negdge_sig;

/************************************************************************
**                           Main Program
************************************************************************/
//对送进来的数据同步处理,通常对单比特的同步处理就是多延几拍,去除采样的亚稳态
always @(posedge sys_clk or negedge sys_rst_n) begin
        if (sys_rst_n == 1'b0) begin
            uart_rxd_dly1 <= 1'b0;              //寄存器清零
            uart_rxd_dly2 <= 1'b0;
            uart_rxd_dly3 <= 1'b0;
            uart_rxd_dly4 <= 1'b0;
        end
        else begin
            uart_rxd_dly1 <= uart_rxd;
            uart_rxd_dly2 <= uart_rxd_dly1;
            uart_rxd_dly3 <= uart_rxd_dly2;
            uart_rxd_dly4 <= uart_rxd_dly3;
        end
end

//下降沿检测,检测到下降沿时对RXD信号采样
assign rxd_negdge_sig = (~uart_rxd_dly3) & uart_rxd_dly4;

//对下降沿做同步打拍处理
```

```verilog
always @(posedge sys_clk or negedge sys_rst_n) begin
        if (sys_rst_n == 1'b0)
            rxd_negdge_sig_dly1 <= 1'b0;            //初始化寄存器
        else
            rxd_negdge_sig_dly1 <= rxd_negdge_sig;

end

//在下降沿等于1且完成一次数据发送计数器清零
always @(posedge sys_clk or negedge sys_rst_n) begin
        if ( sys_rst_n == 1'b0 )
            counter <= 16'b0;           //复位时计数器清零
        else if ( rxd_negdge_sig_dly1 == 1'b1 && counter > 57200 )
            counter <= 16'b0;
        else if ( counter <= 57200 )//完成一次数据接收
            counter <= counter + 16'b1;
        else ;
end

always @(posedge sys_clk or negedge sys_rst_n) begin
    if ( sys_rst_n == 1'b0 )
        buff <= 8'b0;                   //复位清零
    else begin
        case ( counter )
            7800: buff[0] <= uart_rxd_dly4;   //counter == 7800 时对 D[0]采样,存入 buff[0]
            13000: buff[1] <= uart_rxd_dly4;  //counter == 13000 时对 D[1]采样,存入 buff[1]
            18200: buff[2] <= uart_rxd_dly4;  //counter == 18200 时对 D[2]采样,存入 buff[2]
            23400: buff[3] <= uart_rxd_dly4;  //counter == 23400 时对 D[2]采样,存入 buff[3]
            28600: buff[4] <= uart_rxd_dly4;  //counter == 28600 时对 D[2]采样,存入 buff[4]
            33800: buff[5] <= uart_rxd_dly4;  //counter == 33800 时对 D[2]采样,存入 buff[5]
            39000: buff[6] <= uart_rxd_dly4;  //counter == 39000 时对 D[2]采样,存入 buff[6]
            44200: buff[7] <= uart_rxd_dly4;  //counter == 44200 时对 D[2]采样,存入 buff[7]
            default: buff <= buff;      //buff 保持数据
        endcase
    end
end

always @(posedge sys_clk or negedge sys_rst_n) begin
        if (sys_rst_n == 1'b0)
            LED <= 8'b0;                //复位键 LED 全灭
        else
            LED <= buff;                //LED 接收到 PC 端发送的数据并显示
```

end

endmodule
//end of RTL code

3.6.5 实验现象

将开发板与 PC 相连,程序下载到开发板后,再通过串口线或者 USB 串口线与 PC 相连,在上位机软件的图 3.25 所示"字符串输入框"中输入字符串,则在开发板可以看到 LED 表示的相应的 ASCII 码值。

例如,输入数字 9(见图 3.25"字符串输入框"文本框),然后单击"发送"按钮。

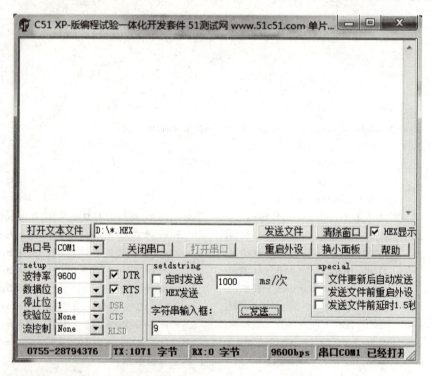

图 3.25 串口接收上位机软件界面

9 对应的 ASCII 码值为 57,开发板上 LED 显示 00111001,如图 3.26 所示。

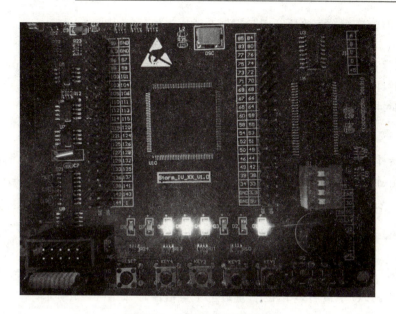

图 3.26 串口接收实验现象

3.7 同步 FIFO 实验

3.7.1 同步 FIFO 简介

FIFO 是一种先进先出的数据缓存器,在逻辑设计里面用的非常多,FIFO 设计可以说是逻辑设计人员必须掌握的常识性设计。FIFO 一般用在隔离两边读/写带宽不一致,或者位宽不一样的地方。

FIFO 包括同步 FIFO 和异步 FIFO 两种。同步 FIFO 有一个时钟信号,读和写逻辑全部使用这一个时钟信号;异步 FIFO 有两个时钟信号,读和写逻辑使用各种读/写时钟。本节说的全部是同步 FIFO。

FIFO 与普通存储器 RAM 的区别是没有外部读/写地址线,使用起来非常简单,但缺点是只能顺序写入数据,顺序读出数据,其数据地址由内部读/写指针自动加 1 完成,不能像普通存储器那样可以由地址线决定读取或写入某个指定的地址。FIFO 本质上是由 RAM 加读/写控制逻辑构成的一种先进先出的数据缓冲器。

1. FIFO 的常见参数

FIFO 的宽度:指 FIFO 一次读/写操作的数据位。

FIFO 的深度:指 FIFO 可以存储多少个 N 位的数据(如果宽度为 N)。

满标志:FIFO 已满或将要满时由 FIFO 的状态电路送出的一个信号,以阻止

FIFO 的写操作继续向 FIFO 中写数据而造成溢出(overflow)。

空标志：FIFO 已空或将要空时由 FIFO 的状态电路送出的一个信号，以阻止 FIFO 的读操作继续从 FIFO 中读出数据而造成无效数据的读出(underflow)。

读时钟：读操作所遵循的时钟，在每个时钟沿来临时读数据。

写时钟：写操作所遵循的时钟，在每个时钟沿来临时写数据。

2. 读/写指针(FIFO 读/写地址)的工作原理

读指针：总是指向下一个将要被写入的单元，复位时，指向第 1 个单元(编号为 0)。

写指针：总是指向当前要被读出的数据，复位时，指向第 1 个单元(编号为 0)。

3. FIFO 的"空"/"满"检测

FIFO 设计的关键：产生可靠的 FIFO 读/写指针和生成 FIFO"空"/"满"状态标志。

当读/写指针相等时，表明 FIFO 为空。这种情况发生在复位操作时，或者当读指针读出 FIFO 中最后一个字后，追赶上了写指针时，如图 3.27 所示。

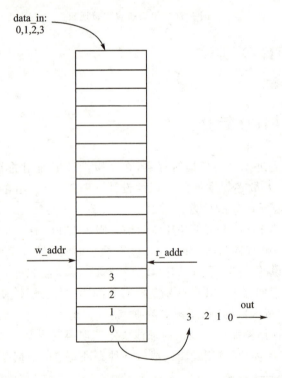

图 3.27 FIFO 读/写地址变化示意图 1

当读/写指针再次相等时，表明 FIFO 为满。这种情况发生在，当写指针转了一圈，折回来(wrapped around)又追上了读指针，如图 3.28 所示。

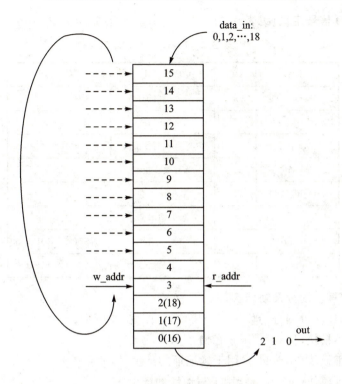

图 3.28　FIFO 读/写地址变化示意图 2

为了区分到底是满状态还是空状态，可以采用以下方法：

在指针中添加一个额外的位(extra bit)，当写指针增加并越过最后一个 FIFO 地址时，就将写指针这个未用的 MSB 加 1，其他位回零。对读指针也进行同样的操作。此时，对于深度为 $2n$ 的 FIFO，需要的读/写指针位宽为 $(n+1)$ 位，如对于深度为 8 的 FIFO，需要采用 4 bit 的计数器，0000～1000、1001～1111，MSB 作为折回标志位，而低 3 位作为地址指针。

如果两个指针的 MSB 不同，说明写指针比读指针多折回了一次；如 r_addr = 0000，而 w_addr = 1000，为满。

如果两个指针的 MSB 相同，则说明两个指针折回的次数相等；如果其余位相等，说明 FIFO 为空。

4. 示　　例

下面设计一个 16×8(16 是深度，8 是位宽)的 FIFO。

(1) 功能定义

用 16×8RAM 实现一个同步先进先出(FIFO)队列设计。由写使能端控制数据流写入 FIFO，并由读使能控制 FIFO 中数据的读出。写入和读出的操作由时钟的上升沿触发。当 FIFO 的数据满和空的时候分别设置相应的高电平加以指示。

(2) FIFO 信号定义(见表 3.2)

表 3.2 FIFO 信号定义

信号名称	I/O	功能描述	备 注
rst	I	全局复位(低有效)	
clk	I	全局时钟	频率 10 MHz
wr_en	I	低有效写使能	
rd_en	I	低有效读使能	
data_in[7:0]	I	数据输入端	
data_out[7:0]	O	数据输出端	
empty	O	空指示信号	为高时表示 FIFO 空
full	O	满指示信号	为高时表示 FIFO 满

(3) 顶层模块划分及功能实现

该同步 FIFO 可划分为如下四个模块,如图 3.29 所示。

① 存储器模块(RAM),用于存储数据;
② 读地址模块(rd_addr),用于读地址的产生;
③ 写地址模块(wr_addr),用于写地址的产生;
④ 标志模块(flag_gen),用于产生 FIFO 当前空满状态。

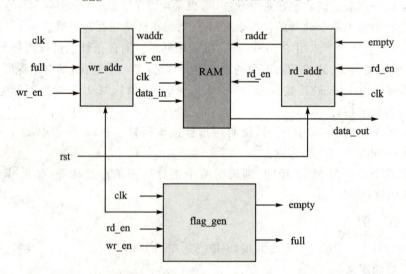

图 3.29 同步 FIFO 的模块划分

第3章 设计实例

1) RAM 模块

本设计中的 FIFO 采用 16×8 双口(一个写口,一个读口)RAM,以循环读/写的方式实现。

读地址:根据 rd_addr 模块产生的读地址,在读使能(rd_en)为高电平的时候,将 RAM 中 rd_addr[3:0]地址中的对应单元的数据在时钟上升沿到来的时候,读出到 data_out[7:0]中。

写地址:根据 wr_addr 产生的写地址,在写使能(wr_en)为高电平的时候,将输入数据(data_in[7:0])在时钟上升沿到来的时候,写入 wr_addr[3:0]地址对应的单元。

2) wr_addr 模块

该模块用于产生 FIFO 写数据时所用的地址。由于 16 深度的 RAM 单元可以用 4 位地址线寻址。本模块用 4 位计数器(wr_addr[3:0])实现写地址的产生。

在复位信号为 0(复位有效)时,写地址值为 0。

如果 FIFO 未满(full=0,FIFO 满了如果还继续写入,会导致新数据覆盖原来的旧数据)且有写使能(wr_en)有效,则 wr_addr[3:0]加 1;否则写地址不变。

3) rd_addr 模块

该模块用于产生 FIFO 读数据时所用的地址。由于 16 个 RAM 单元可以用4 位地址线寻址。本模块用 4 位计数器(rd_addr[3:0])实现读地址的产生。

在复位信号为 0(复位有效)时,读地址值为 0。

如果 FIFO 未空(empty=0,FIFO 空了如果还继续读,会导致 FIFO 读出的数据不是期望的数据或者不确定的数据)且有读使能(rd_en)有效,则 rd_addr[3:0]加 1;否则读地址不变。

4) flag_gen 模块

flag_gen 模块产生 FIFO 空满标志。本模块设计没有使用读/写地址判定 FIFO 是否空满(也可以使用读/写地址做判断),而是设计一个计数器,该计数器(ptr_cnt)用于指示当前 FIFO 中数据的个数。由于 FIFO 中最多只有 16 个数据,因此采用 5 位计数器来指示 FIFO 中数据的个数。具体计算如下:

复位的时候,ptr_cnt=0;

如果 wr_en 和 rd_en 同时有效的时候,则 ptr_cnt 不加也不减,表示同时对 FIFO 进行读/写操作的时候,FIFO 中的数据个数不变。

如果 wr_en 有效且 full=0,则 pt_cont+1,表示写操作且 FIFO 未满的时候,FIFO 中的数据个数增加了 1。

如果 rd_en 有效且 empty=0,则 pt_cont-1,表示读操作且 FIFO 未满的时候,FIFO 中的数据个数减少了 1。

如果 ptr_cnt=0,表示 FIFO 空,需要设置 empty=1;如果 ptr_cnt=16,表示 FIFO 现在已经满,需要设置 full=1。

3.7.2 实验任务

设计一个 FIFO,具有读/写控制逻辑、空满指示信号。

3.7.3 硬件设计

该实验不适合做硬件实现,适合 Modelsim 软件仿真,所以不涉及硬件。

3.7.4 程序设计

代码如下:

```verilog
module sync_fifo(
input              clk,
input              rst,
input              wr_en,
input              rd_en,
input     [7:0]    data_in,
output reg [7:0]   data_out,
output reg         empty,
output reg         full
);

reg [3:0]  wr_addr;
reg [3:0]  rd_addr;
reg [4:0]  count;
parameter max_count = 5'b10000 ;
parameter max1_count = 5'b01111 ;
reg [7:0]  fifo    [0:max_count];      //定义一个二维的 RAM,实际设计中根据 RAM 深度
                                       //和宽度可能会采用 Vedor 提供的 RAM

//读操作
always @ (posedge clk or negedge rst) begin
    if (rst == 1'b0)
      data_out <= 0;
    else if (rd_en && empty == 0)
      data_out <= fifo[rd_addr];
  end
//写操作
```

```verilog
always @ (posedge clk ) begin
    if (wr_en == 1&&full == 0)
        fifo[wr_addr]<  = data_in;
end
//更新读地址
always @ (posedge clk or negedge rst) begin
    if (rst ==  1'b0)
        rd_addr< = 4'b0000;
    else if (empty == 0&&rd_en == 1)
        rd_addr< = rd_addr + 1;
end
//更新写地址
always @ (posedge clk or negedge rst) begin
    if (rst ==  1'b0)
        wr_addr< = 4'b0000;
    else if (full == 0&&wr_en == 1)
        wr_addr< = wr_addr + 1;
    else
        wr_addr< = 4'b0000;
end
//更新标志位
always @ (posedge clk or negedge rst) begin
    if (rst ==  1'b0)
     count< = 0;
    else begin
     case({wr_en,rd_en})
     2'b00:count< = count;
     2'b01:
            if(count! == 5'b00000)
              count< = count - 1;
     2'b10:
            if(count! == max1_count)
              count< = count + 1;
     2'b11:count< = count;
     endcase
    end
end

always @(count) begin
    if(count == 5'b00000)
      empty< = 1;
    else
```

```verilog
        empty <= 0;
    end

    always @(count) begin
        if (count == max_count)
            full <= 1;
        else
            full <= 0;
    end
endmodule
```

下面是 TestBench 代码(使用 Modelsim 仿真使用的激励代码):

TestBench.v:

```verilog
`timescale 1 ns/ 1 ps
module sync_fifo_vlg_tst();

reg             clk;
reg     [7:0]   data_in;
reg             rd_en;
reg             rst;
reg             wr_en;

// wires
wire    [7:0]   data_out;
wire            empty;
wire            full;
reg     [7:0]   dindely;

sync_fifo       i1 (
    .clk        (clk),
    .data_in    (data_in),
    .data_out   (data_out),
    .empty      (empty),
    .full       (full),
    .rd_en      (rd_en),
    .rst        (rst),
    .wr_en      (wr_en)
);

always @(posedge clk) begin
    dindely <= data_in;
end
```

```
initial begin
    clk = 0;
    rst = 0;
    rd_en = 0;
    wr_en = 0;
    data_in = 0;

    #40
    rst = 1;
    #35
    wr_en = 1;
    #400
    rd_en = 1;
end

always #20 data_in <= data_in + 1;

always #10 clk = ~clk;

endmodule
```

3.7.5 实验现象

本实验不适合硬件实现,适合 Modelsim 软件仿真,仿真波形如图 3.30 所示。

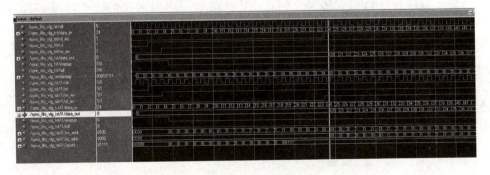

图 3.30 同步 FIFO 实验读/写仿真波形

3.8 异步 FIFO 实验

3.8.1 异步 FIFO 简介

FIFO 包括同步 FIFO 和异步 FIFO 两种。同步 FIFO 有一个时钟信号，读和写逻辑全部使用这一个时钟信号；异步 FIFO 有两个时钟信号，读和写逻辑用各种读/写时钟。本节说的全部是异步 FIFO。

1. 异步 FIFO 的用途

异步 FIFO 读/写分别采用不同时钟。这两个时钟可能时钟频率不同，也可能时钟相位不同，可能是同源时钟，也可能是不同源时钟。

在现代逻辑设计中，随着设计规模的不断扩大，一个系统中往往含有数个时钟，多时钟域带来的一个问题就是，如何设计异步时钟之间的接口电路。异步 FIFO 是这个问题的一种简便、快捷的解决方案，使用异步 FIFO 可以在两个不同时钟系统之间快速而方便地传输实时数据。

2. FIFO 的常见参数

FIFO 的宽度：指 FIFO 一次读/写操作的数据位。

FIFO 的深度：指 FIFO 可以存储多少个 N 位的数据（如果宽度为 N）。

满标志：FIFO 已满或将要满时由 FIFO 的状态电路送出的一个信号，以阻止 FIFO 的写操作继续向 FIFO 中写数据而造成溢出（overflow）。

空标志：FIFO 已空或将要空时由 FIFO 的状态电路送出的一个信号，以阻止 FIFO 的读操作继续从 FIFO 中读出数据而造成无效数据的读出（underflow）。

读时钟：读操作所遵循的时钟，在每个时钟沿来临时读数据。

写时钟：写操作所遵循的时钟，在每个时钟沿来临时写数据。

3. 读/写指针(FIFO 读/写地址)的工作原理

读指针：总是指向下一个将要被写入的单元，复位时，指向第 1 个单元（编号为 0）。

写指针：总是指向当前要被读出的数据，复位时，指向第 1 个单元（编号为 0）。

FIFO 的"空"/"满"检测

FIFO 设计的关键：产生可靠的 FIFO 读/写指针和生成 FIFO"空"/"满"状态标志。

当读/写指针相等时，表明 FIFO 为空。这种情况发生在复位操作时，或者当读指针读出 FIFO 中最后一个字后，追赶上了写指针时，如图 3.31 所示。

当读写指针再次相等时，表明 FIFO 为满。这种情况发生在，当写指针转了一圈，折回来（wrapped around）又追上了读指针，如图 3.32 所示。

第 3 章 设计实例

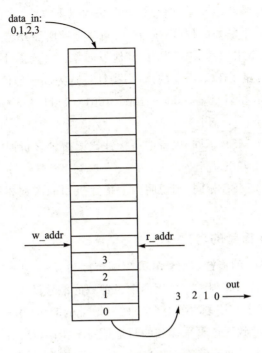

图 3.31 FIFO 读/写地址变化示意图 1

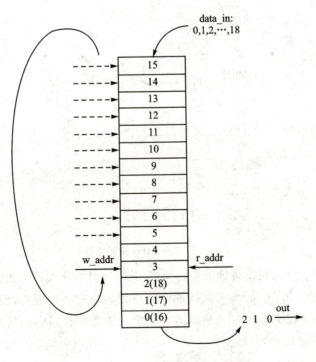

图 3.32 FIFO 读/写地址变化示意图 1

为了区分到底是满状态还是空状态,可以采用以下方法:

在指针中添加一个额外的位(extra bit),当写指针增加并越过最后一个FIFO地址时,就将写指针这个未用的MSB加1,其他位回零。对读指针也进行同样的操作。此时,对于深度为$2n$的FIFO,需要的读/写指针位宽为$(n+1)$位,如对于深度为8的FIFO,需要采用4 bit的计数器,0000~1000、1001~1111,MSB作为折回标志位,而低3位作为地址指针。

如果两个指针的MSB不同,则说明写指针比读指针多折回了一次,如r_addr=0000,而w_addr=1000,为满。

如果两个指针的MSB相同,则说明两个指针折回的次数相等。其余位相等,说明FIFO为空。

4. 异步FIFO指针的考虑

为什么异步FIFO的指针需要特殊处理呢?因为异步FIFO的空满指示需要使用指针进行判断,如果空满判断直接用两个时钟域的信号做逻辑判断,会导致逻辑判断错误,如图3.33所示。读逻辑的时钟采样写地址的时候,由于路径延迟和两个时钟相位都不同,就会导致读逻辑采样到的写地址完全错误(如果采样的时候刚好写地址信号跳变,还可能会导致亚稳态产生),导致空标记错误,导致FIFO不能使用。

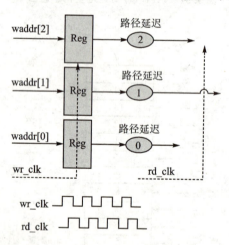

图3.33 异步FIFO指针变化

一般异步FIFO的地址传递需要使用格雷码(一种数字编码格式,gray code),格雷码转换结构如图3.34所示,每次地址变化格雷码只有一位变化,因此在两个时钟域间同步多个位不会产生问题。所以需要一个二进制到格雷码的转换电路,将地址值转换为相应的格雷码,然后将该格雷码同步到另一个时钟域进行对比,作为空满状态的检测。

第 3 章 设计实例

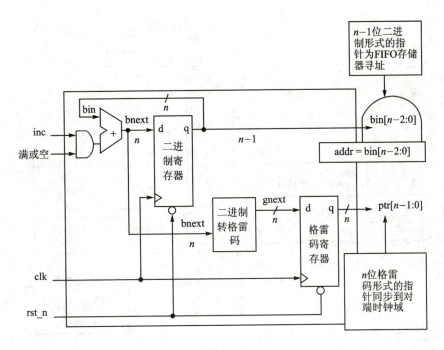

图 3.34 异步 FIFO 格雷码

使用格雷码进行对比,如何判断"空"与"满"?

使用格雷码解决了指针采样错误的问题,但同时也带来另一个问题,即在格雷码域如何判断空与满。

对于"空"的判断依然依据二者完全相等(包括 MSB)。

而对于"满"的判断,如图 3.35 所示,由于格雷码除了 MSB 外,具有镜像对称的特点,当读指针指向 7,写指针指向 8 时,除了 MSB,其余位皆相同,不能说它为满。因此不能单纯地只检测最高位了,在格雷码上判断为满必须同时满足以下 3 个条件:

- wptr 和同步过来的 rptr 的 MSB 不相等,因为 wptr 必须比 rptr 多折回一次。
- wptr 与 rptr 的次高位不相等,如图 3.35 所示位置 7 和位置 15,转化为二进制对应的是 0111 和 1111,MSB 不同说明多折回一次,相同(111)代表同一位置。
- 剩下的其余位完全相等。

异步 FIFO 总体框图如 3.36 所示。主要包括写指针和满信号产生、读指针和空信号产生、时钟域转换逻辑。

3.8.2 实验任务

设计一个异步 FIFO,具有读/写控制逻辑、空满指示信号。

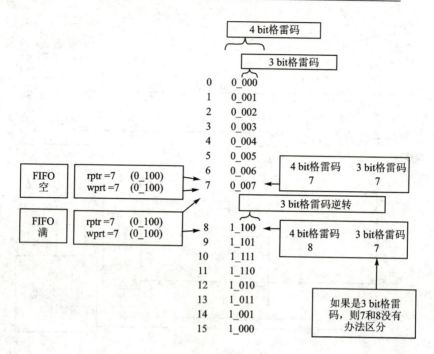

图 3.35 异步 FIFO 空满状态

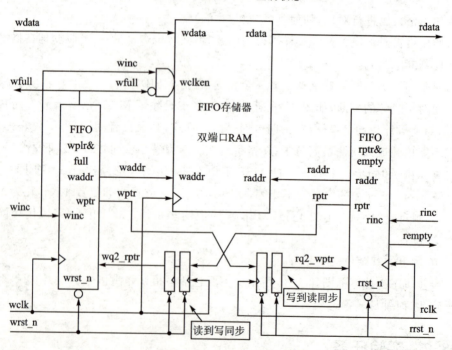

图 3.36 异步 FIFO 总体框图

3.8.3 硬件设计

本实验不适合做硬件实现，适合 Modelsim 软件仿真，所以不涉及硬件。

3.8.4 程序设计

代码如下：

```verilog
module AsyncFIFO
#(parameter ASIZE = 4,         //地址位宽
  parameter DSIZE = 8)         //数据位宽
 (
    input  [DSIZE-1:0] wdata,
    input              winc, wclk, wrst_n,   //写请求信号,写时钟,写复位
    input              rinc, rclk, rrst_n,   //读请求信号,读时钟,读复位
    output [DSIZE-1:0] rdata,
    output             wfull,
    output             rempty
 );
wire [ASIZE-1:0] waddr;
wire [ASIZE-1:0] raddr;
wire [ASIZE:0]   wptr;
wire [ASIZE:0]   rptr;
wire [ASIZE:0]   wq2_rptr;
wire [ASIZE:0]   rq2_wptr;

/***********************************************************
* In order to perform FIFO full and FIFO empty tests using
* this FIFO style, the read and write pointers must be
* passed to the opposite clock domain for pointer comparison
***********************************************************/
//在检测"满"或"空"状态之前,需要将指针同步到其他时钟域时,使用格雷码,
//可以降低同步过程中亚稳态出现的概率
sync_r2w I1_sync_r2w(
    .wq2_rptr(wq2_rptr),
    .rptr(rptr),
    .wclk(wclk),
    .wrst_n(wrst_n));
sync_w2r I2_sync_w2r (
    .rq2_wptr(rq2_wptr),
```

```
        .wptr(wptr),
        .rclk(rclk),
        .rrst_n(rrst_n));
    /*
     *   DualRAM
     */
    DualRAM #(DSIZE, ASIZE) I3_DualRAM(
        .rdata(rdata),
        .wdata(wdata),
        .waddr(waddr),
        .raddr(raddr),
        .wclken(winc),
        .wclk(wclk));
    /*
     * 空、满比较逻辑
     */
    rptr_empty #(ASIZE) I4_rptr_empty(
        .rempty(rempty),
        .raddr(raddr),
        .rptr(rptr),
        .rq2_wptr(rq2_wptr),
        .rinc(rinc),
        .rclk(rclk),
        .rrst_n(rrst_n));
    wptr_full #(ASIZE) I5_wptr_full(
        .wfull(wfull),
        .waddr(waddr),
        .wptr(wptr),
        .wq2_rptr(wq2_rptr),
        .winc(winc),
        .wclk(wclk),
        .wrst_n(wrst_n));
endmodule
```

DualRAM 模块：

```
module DualRAM
#(
    parameter DATA_SIZE = 8,      //数据位宽
    parameter ADDR_SIZE = 4       //地址位宽
)
(
    input                 wclken,wclk,
```

```verilog
    input   [ADDR_SIZE-1:0] raddr,      //RAM read address
    input   [ADDR_SIZE-1:0] waddr,      //RAM write address
    input   [DATA_SIZE-1:0] wdata,      //data input
    output  [DATA_SIZE-1:0] rdata       //data output
);
localparam RAM_DEPTH = 1 << ADDR_SIZE;  //RAM 深度 = 2^ADDR_WIDTH
        reg [DATA_SIZE-1:0] Mem[RAM_DEPTH-1:0];
        always@(posedge wclk)
begin
    if(wclken)
        Mem[waddr] <= wdata;
end
assign rdata = Mem[raddr];
endmodule
```

同步模块 1(按照图 3.36 设计,此时指针经过两个寄存器,做打二拍同步(使用 2 个寄存器寄存同步)处理):

```verilog
module sync_r2w
#(parameter ADDRSIZE = 4)
(
    output reg  [ADDRSIZE:0]    wq2_rptr,
    input       [ADDRSIZE:0]    rptr,
    input                       wclk, wrst_n
);
reg [ADDRSIZE:0] wq1_rptr;
always @(posedge wclk or negedge wrst_n)
    if(!wrst_n)
        {wq2_rptr,wq1_rptr} <= 0;
    else
        {wq2_rptr,wq1_rptr} <= {wq1_rptr,rptr};
endmodule
```

同步模块 2:

```verilog
module sync_w2r
#(parameter ADDRSIZE = 4)
(
    output reg  [ADDRSIZE:0]    rq2_wptr,
    input       [ADDRSIZE:0]    wptr,
    input                       rclk, rrst_n
);  reg [ADDRSIZE:0] rq1_wptr;
always @(posedge rclk or negedge rrst_n)
    if(!rrst_n)
```

```
        {rq2_wptr,rq1_wptr} <= 0;
    else
        {rq2_wptr,rq1_wptr} <= {rq1_wptr,wptr};
endmodule
```

空判断逻辑:

```
module rptr_empty
#(parameter ADDRSIZE = 4)
(
    output reg     rempty,
    output         [ADDRSIZE-1:0]   raddr,
    output reg     [ADDRSIZE:0]     rptr,
    input          [ADDRSIZE:0]     rq2_wptr,
    input          rinc, rclk, rrst_n);


reg [ADDRSIZE:0] rbin;
wire [ADDRSIZE:0] rgraynext, rbinnext;
wire rempty_val;
//------------------
// GRAYSTYLE2 pointer:格雷码读地址指针
//------------------
always @(posedge rclk or negedge rrst_n)
    if(!rrst_n)
        begin
            rbin <= 0;
            rptr <= 0;
        end
    else
        begin
            rbin <= rbinnext;
            rptr <= rgraynext;
        end
// 格雷码计数逻辑
assign rbinnext = !rempty ? (rbin + rinc) : rbin;
assign rgraynext = (rbinnext>>1) ^ rbinnext;         //二进制到格雷码的转换
    assign raddr = rbin[ADDRSIZE-1:0];
//------------------------------------
// FIFO empty when the next rptr == synchronized wptr or on reset
//------------------------------------
/*
 *读指针是一个 n 位的格雷码计数器,比 FIFO 寻址所需的位宽大一位
```

*当读指针和同步过来的写指针完全相等时(包括 MSB),说明二者折回次数一致,FIFO 为空
*/
assign rempty_val = (rgraynext == rq2_wptr);
        always @(posedge rclk or negedge rrst_n)
if (! rrst_n)
    rempty <= 1'b1;
else
    rempty <= rempty_val;
endmodule

满判断逻辑:

module wptr_full
    #(
    parameter ADDRSIZE = 4
)
    (
    output reg   wfull,
    output   [ADDRSIZE-1:0]   waddr,
    output reg   [ADDRSIZE:0]   wptr,
    input   [ADDRSIZE:0]   wq2_rptr,
    input   winc, wclk, wrst_n);

reg [ADDRSIZE:0] wbin;
wire [ADDRSIZE:0] wgraynext, wbinnext;
wire wfull_val;
// GRAYSTYLE2 pointer
always @(posedge wclk or negedge wrst_n)
    if (! wrst_n)
    begin
        wbin <= 0;
        wptr <= 0;
    end
    else
    begin
        wbin <= wbinnext;
        wptr <= wgraynext;
    end
//gray 码计数逻辑
assign wbinnext = ! wfull ? wbin + winc : wbin;
assign wgraynext = (wbinnext>>1) ^ wbinnext;
        assign waddr = wbin[ADDRSIZE-1:0];
        /* 由于满标志在写时钟域产生,因此比较安全的做法是将读指针同步到写时钟域 */

```
//--------------------------------------------------
// Simplified version of the three necessary full-tests:
// assign wfull_val = ((wgnext[ADDRSIZE] != wq2_rptr[ADDRSIZE]) &&
// (wgnext[ADDRSIZE-1] != wq2_rptr[ADDRSIZE-1]) &&
// (wgnext[ADDRSIZE-2:0] == wq2_rptr[ADDRSIZE-2:0]));
//--------------------------------------------------
assign wfull_val = (wgraynext == {~wq2_rptr[ADDRSIZE:ADDRSIZE-1],
                                   wq2_rptr[ADDRSIZE-2:0]});
always @(posedge wclk or negedge wrst_n)
    if (!wrst_n)
        wfull <= 1'b0;
    else
        wfull <= wfull_val;
endmodule
```

注意：

要判断当前 FIFO 的"空""满"状态时往往是把写地址同步到读时钟进行"空"的判断，把读地址同步到写时钟进行"满"判断。试想一下，为什么要这样？这是因为"空"和"满"的判断是立即有效的，如果把读地址同步到写时钟去判断"空"标志，那么得到的有效"空"标志是在写时钟上的；要把"空"发送到外部的读逻辑，则需要先同步到读时钟上。这样就要经过 2 个锁存器，延时 2 个时钟周期；而在这 2 个时钟周期上，读操作可能已经多读了。和"空""满"标志的立即有效相比，"空""满"标志的失效是延时 2 个时钟的；因为读/写地址都需要经过二级锁存器同步之后才会进行判断。同样的道理，由于有二级锁存器的延时，FIFO 的"空""满"不一定是真的"空""满"，写地址经过 2 个时钟延时同步到读时钟进行"空"判断。如果"空"有效，并不意味着 FIFO 一定是"空"了，有可能写逻辑在这 2 个时钟的延时中写进了新的数据，如果写地址没有变化那 FIFO 就是读"空"了。

TESTBENCH 测试平台代码：

```
`timescale 1 ns/ 1 ps
module AsyncFIFO_vlg_tst();
reg rclk;
reg rinc;
reg rrst_n;
reg wclk;
reg [7:0] wdata;
reg winc;
reg wrst_n;
wire [7:0] rdata;
wire rempty;
```

```verilog
wire wfull;
AsyncFIFO i1 (
    .rclk(rclk),
    .rdata(rdata),
    .rempty(rempty),
    .rinc(rinc),
    .rrst_n(rrst_n),
    .wclk(wclk),
    .wdata(wdata),
    .wfull(wfull),
    .winc(winc),
    .wrst_n(wrst_n)
);
initial
begin
    rrst_n = 0;
    wrst_n = 0;
    rclk = 0;
    wclk = 0;
    winc = 0;
    rinc = 0;
    #20 rrst_n = 1;
    wrst_n = 1 ;
end

initial begin
    rclk = 'b0;
    wclk = 'b0;

end

always #10 rclk = ~ rclk;
always #10 wclk = ~ wclk;

always @(posedge rclk)
    rinc <= {$random} % 2;
always @(posedge wclk)
    winc <= {$random} % 2;
```

```
always @(negedge wclk)
    wdata <= {$random} % 256;
endmodule
```

3.8.5 实验现象

本实验不适合做硬件实现,适合 Modelsim 软件仿真,仿真波形如图 3.37 所示。

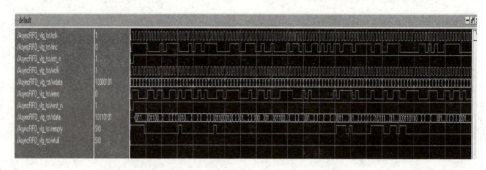

图 3.37 异步 FIFO 实验读/写仿真波形图

3.9 状态机实验

3.9.1 状态机简介

有限状态机(Finite State Machine),简称状态机,缩写为 FSM。

有限状态机是指输出取决于过去输入部分和当前输入部分的时序逻辑电路。有限状态机又可以认为是组合逻辑和寄存器逻辑的一种组合。状态机特别适合描述那些发生有先后顺序或者有逻辑规律的事情,其实这就是状态机的本质。状态机就是对具有逻辑顺序或时序规律的事件进行描述的一种方法。

根据状态机的输出是否与输入条件相关,可将状态机分为两大类,即摩尔(Moore)型状态机和米勒(Mealy)型状态机。

- Mealy 状态机:时序逻辑的输出不仅取决于当前状态,还取决于输入。
- Moore 状态机:时序逻辑的输出只取决于当前状态。

根据实际写法,状态机还可以分为一段式、二段式和三段式状态机。

- 一段式:把整个状态机写在一个 always 模块中,并且这个模块既包含状态转移,又含有组合逻辑输入/输出。
- 二段式:状态切换用时序逻辑,次态输出和信号输出用组合逻辑。

- 三段式:状态切换用时序逻辑,次态输出用组合逻辑,信号输出用时序逻辑。

1. Mealy 状态机(见图 3.38)

- 下一个状态 = F(当前状态,输入信号);
- 输出信号 = G(当前状态,输入信号)。

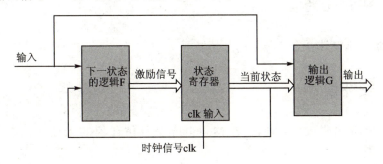

图 3.38 Mealy 状态机

2. Moore 状态机(见图 3.39)

- 下一个状态 = F(当前状态,输入信号);
- 输出信号 = G(当前状态)。

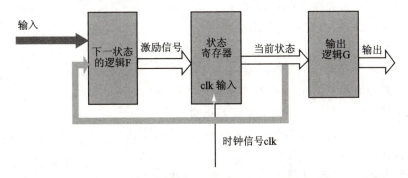

图 3.39 Moore 状态机

3. 三段式状态机

逻辑设计里面,最常用的是二段式和三段式的状态机。三段式状态机比二段式状态机更好,下面重点介绍三段式状态机。

二段式直接采用组合逻辑输出,而三段式则通过在组合逻辑后再增加一级寄存器来实现时序逻辑输出。这样做的好处是可以有效滤去组合逻辑输出的毛刺,同时可以有效地进行时序计算与约束。另外,对于总线形式的输出信号来说,容易使总线数据对齐,从而减小总线数据间的偏移,减小接收端数据采样出错的频率。

三段式状态机的基本格式如下:

- 第一个 always 语句实现同步状态跳转；
- 第二个 always 语句实现组合逻辑；
- 第三个 always 语句实现同步输出。

3.9.2 实验任务

使用状态机设计一个 7 分频的分频器。

3.9.3 硬件设计

本实验不适合做硬件实现，适合 Modelsim 软件仿真，不涉及硬件。

3.9.4 程序设计

1. 设计思路

描述状态机需要注意的事项：
- 定义模块名和输入/输出端口；
- 定义输入/输出变量或寄存器；
- 定义时钟和复位信号；
- 定义状态变量和状态寄存器；
- 用时钟沿触发的 always 块表示状态转移过程；
- 在复位信号有效时给状态寄存器赋初始值；
- 描述状态的转换过程：符合条件，从一个状态到另外一个状态，否则留在原状态；
- 验证状态转移的正确性，必须完整和全面。

2. 方案说明

① 本状态机采用独热码设计，简称 one-hot code，独热码编码的最大优势在于状态比较时仅仅需要比较一个位，从而一定程度上简化了译码逻辑。

② 一般状态机状态编码使用二进制编码（binary）、格雷码（gray-code）、独热码（one-hot code）。

3. 各种编码比较

二进制编码、格雷码编码使用最少的触发器，消耗较多的组合逻辑，而独热码编码反之。独热码编码的最大优势在于状态比较时仅仅需要比较一个位，从而一定程度上简化了译码逻辑。虽然在需要表示同样的状态数时，独热编码占用较多的位，也就是消耗较多的触发器,但这些额外触发器占用的面积可与译码电路省下来的面积相抵消。

binary(二进制编码)、gray-code(格雷码)编码使用最少的触发器,较多的组合逻辑,而 one-hot(独热码)编码反之。one-hot 编码的最大优势在于状态比较时仅仅需要比较一个 bit,一定程度上从而简化了比较逻辑,减少了毛刺产生的概率。另一方面,对于小型设计使用 gray-code 和 binary 编码更有效,而大型状态机使用 one-hot 码更高效。

4. 源代码

```verilog
module divider7_fsm (
//input
input       sys_clk,        // system clock;
input       sys_rst_n,      // system reset, low is active;

//output
output reg  clk_divide_7    // output divide 7 clk
            );

//reg define
reg [6:0]   curr_st;        // FSM current state
reg [6:0]   next_st;        // FSM next state
reg         clk_divide_7;   // generated clock,divide by 7


//wire define

//parameter define

//one hot code design
parameter S0 = 7'b0000000;
parameter S1 = 7'b0000001;
parameter S2 = 7'b0000010;
parameter S3 = 7'b0000100;
parameter S4 = 7'b0001000;
parameter S5 = 7'b0010000;
parameter S6 = 7'b0100000;


/***************************************
**                Main Program
***************************************/
//generate FSM next state
```

```verilog
always @(posedge sys_clk or negedge sys_rst_n) begin
        if (sys_rst_n == 1'b0) begin
            curr_st <= 7'b0;
        end
        else  begin
            curr_st <= next_st;
        end
end

//FSM state logic
always @( * ) begin
    case (curr_st)
        S0: begin
                next_st = S1;
            end
        S1: begin
                next_st = S2;
            end
        S2: begin
                next_st = S3;
            end
        S3: begin
                next_st = S4;
            end
        S4: begin
                next_st = S5;
            end
        S5: begin
                next_st = S6;
            end
        S6: begin
                next_st = S0;
            end
        default: next_st = S0;
    endcase
end

//control divide clock offset
always @(posedge sys_clk or negedge sys_rst_n) begin
        if (sys_rst_n == 1'b0) begin
            clk_divide_7 <= 1'b0;
        end
```

```
        else if((curr_st == S0) | (curr_st == S1) | (curr_st == S2)| (curr_st == S3))
            clk_divide_7 <= 1'b0;
        else if((curr_st == S4) | (curr_st == S5) | (curr_st == S6))
            clk_divide_7 <= 1'b1;
        else
            ;
end

endmodule
//end of RTL code
```

3.9.5 实验现象

使用 Quratus Ⅱ 的 SignalTap Ⅱ 抓取到的波形如图 3.40 所示，可以看出波形为 7 分频后的时钟信号。

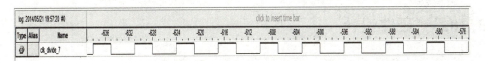

图 3.40　状态机 SignalTap Ⅱ 波形

3.10　EEPROM 写操作实验

3.10.1　EEPROM 写操作简介

EEPROM 是可在线电擦除和电写入的存储器，具有体积小、接口简单、数据保存可靠、可在线改写、功耗低等特点，而且为低电压写入，在电子系统中应用十分普遍。一般 EEPROM 都是串行接口。

串行 EEPROM 按总线形式分为三种，即 I2C 总线、Microwire 总线及 SPI 总线三种。本文将以 Atmel 公司的 I2C 接口 EEPROM 芯片为例进行介绍。

- 工作电压范围 1.8～6.0 V；
- 低功耗 CMOS 技术；
- 写保护功能，当 WP 为高电平时进入写保护状态；
- 页写缓冲器；
- 自定时擦写周期；
- 1 000 000 编程/擦除周期；

- 可保存数据 100 年；
- 8 引脚 DIP SOIC 或 TSSOP 封装。

本实验的 EEPROM 采用 Atmel 公司的 AT24C01/02/04/08/16 系列的 AT24C16 EEPROM。这几种 EEPROM 操作时序是一样的，不同的数字代表容量的差异。

AT24C01/02/04/08/16 是一个 1K/2K/4K/8K/16K 位串行 CMOS EEPROM 内部含有 128/256/512/1 024/2 048 个 8 位字节。

AT24C01 有一个 8 字节页写缓冲器，AT24C02/04/08/16 有一个 16 字节页写缓冲器，该器件通过 I2C 总线接口进行操作有一个专门的写保护功能。

3.10.2 实验任务

EEPROM 的写操作主要任务是设计 EEPROM 的写时序，进行某个地址的数据写入过程。

3.10.3 硬件设计

EEPROM 写操作实验原理图如图 3.41 所示。

A1/A2/A3 地址代表每个器件的编号，开发板上三个地址位全部接地，因此 device_addr = 3'b000（程序中会用到）。

WP 接地，处于写保护关闭状态。

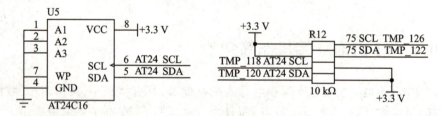

图 3.41　EEPROM 写操作实验原理图

I2C 接口需要接上拉电阻，原因是 I2C 接口是漏极开路模式，需要接上拉电阻拉到确定的电平上。

3.10.4 程序设计

1. 设计思路

本次实验也是主要根据 AT24CXX 的字节写时序要求以及 I2C 总线协议的规定

来编写程序。

AT24CXX 时钟频率如表 3.3 所列。

表 3.3　AT24CXX 时钟频率参数

符　号	参　数	1.8 V,2.5 V		4.5～5.5 V		单　位
		最小	最大	最小	最大	
F_{SCL}	时钟频率		100		400	kHz
T_I	SCL、SDA 输入的噪声抑制时间		200		200	ns

由于系统时钟为 50 MHz，为了分频方便，此处选用 50 kHz 的频率作为 AT24C16 时钟频率。

分频计数器最大需要计数到 50 MHz/50 kHz ＝ 50 000 000/50 000 ＝ 1 000。

AT24CXX 的写操作包括：字节写和页写，本文采用字节写模式。

2. 字节写

在字节写模式下主器件发送起始命令和从器件地址信息 R/W 位置零给从器件，在从器件产生应答信号后，主器件发送 AT24CXX 的字节地址，主器件在收到从器件的另一个应答信号后，再发送数据到被寻址的存储单元，AT24CXX 再次应答，并在主器件产生停止信号后开始内部数据的擦写。在内部擦写过程中 AT24CXX 不再应答主器件的任何请求。

AT24CXX 字节写时序如图 3.42 所示。

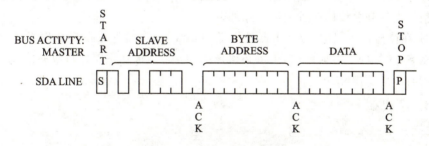

图 3.42　AT24CXX 字节写时序

时序主要包括起始位、从器件地址、字节地址、数据、ACK 和 STOP 位。一次字节写操作总共需要 29 个 I2C 时钟周期，本设计使用 SCL 的四倍频率来控制计数器，4×29 小于 256，所以计数器宽度为 8 位即可。本设计使用计数器控制 SDA 相对于 SCL 的相位关系（开始和结束就是通过相位调整来得到的）。

3. I2C 总线数据通信协议规范（必须遵守）

主机和 DEVICE（设备）之间的通信必须严格遵循 I2C 总线管理定义的规则。DEVICE（设备）寄存器读/写操作的协议通过下列步骤说明：

① 通信开始之前，I2C 总线必须空闲或者不忙。这就意味着总线上的所有器件

都必须释放 SCL 和 SDA 线,SCL 和 SDA 线被总线的上拉电阻拉高。

② 由主机来提供通信所需的 SCL 时钟脉冲。在连续的 9 个 SCL 时钟脉冲作用下,数据(8 位的数据字节以及紧跟其后的 1 个应答状态位)被传输。

③ 在数据传输过程中,除起始和停止信号外,SDA 信号必须保持稳定,而 SCL 信号必须为高。这就表明 SDA 信号只能在 SCL 为低时改变。

④ S:起始信号,主机启动一次通信的信号,SCL 为高电平,SDA 从高电平变成低电平。

⑤ RS:重复起始信号,与起始信号相同,用来启动一个写命令后的读命令。

⑥ P:停止信号,主机停止一次通信的信号,SCL 为高电平,SDA 从低电平变成高电平。然后总线变成空闲状态。

⑦ W:写位,在写命令中写/读位=0。

⑧ R:读位,在读命令中写/读位=1。

⑨ A:器件应答位,由 DEVICE(设备)返回。当器件正确工作时该位为 0,否则为 1。为了使器件获得 SDA 的控制权,这段时间内主机必须释放 SDA 线。

⑩ A':主机应答位,不是由器件返回,而是在读 2 字节的数据时由主控器或主机设置的。在这个时钟周期内,为了告知器件的第一个字节已经读走并要求器件将第二个字节放到总线上,主机必须将 SDA 线设为低电平。

⑪ NA:非应答位。在这个时钟周期内,数据传输结束时器件和主机都必须释放 SDA 线,然后由主机产生停止信号。

⑫ 在写操作协议中,数据从主机发送到器件,由主机控制 SDA 线,但在器件将应答信号发送到总线的时钟周期内除外。

⑬ 在读操作协议中,数据由器件发送到总线上,在器件正在将数据发送到总线和控制 SDA 线的这段时间内,主机必须释放 SDA 线,但在主器件将应答信号发送到总线的时间周期内除外。

4. 源代码

```
module e2prom_write (
    input           sys_clk,            //system clock
    input           sys_rst_n,          //system reset, low is active

    inout           i2c_sda,

    //output ports
    output          i2c_sclk,

    output reg [7:0]    LED
);
```

```
//parameter define
parameter    WIDTH = 8;

//reg define
reg     [WIDTH - 1:0]           counter;
reg     [9:0]                   counter_div;

reg                             clk_50k;
reg                             clk_200k;

reg                             i2c_sclk;
reg                             sda;

reg                             enable;
reg     [WIDTH - 1:0]           data_out;

reg     [31:0]                  counter_init;
//wire define

wire    [2:0]                   device_addr;
wire    [7:0]                   memory_addr;

wire                            sda_input;
wire    [10:0]                  buff;


/**************************************************
**                      Main Program
**************************************************/

//AT24C16 device address is 000, this value is no care
assign device_addr = 3'b000;        //device address is 000

// AT24C16 write addr
assign memory_addr = 8'h02;         //memeory address is 02

// AT24C16 write data
assign buff = 8'h55;                //memeory data is 8'h55

// counter for gen a clk_50k : need count to 1 000, for 50 MHz/1 000 = 50 kHz
always @(posedge sys_clk or negedge sys_rst_n) begin
        if (sys_rst_n == 1'b0)
```

```verilog
                counter_div <= 10'b0;
        else if (counter_div >= 10'd999)
                counter_div <= 10'b0;
        else
                counter_div <= counter_div + 10'b1;
end


// gen a clk_50k use counter_div ;not use counter_div 0 - 500 is for i2c bus request
// start timing
always @(posedge sys_clk or negedge sys_rst_n) begin
        if (sys_rst_n == 1'b0)
                clk_50k <= 10'b0;
        else if ((counter_div >= 375) && (counter_div < 875))
                clk_50k <= 10'b1;
        else
                clk_50k <= 10'b0;
end


// counter for init for AT24C16
//延迟一定时间(时间具体值没有意义),EEPROM 内部初始化
always @(posedge sys_clk or negedge sys_rst_n) begin
        if (sys_rst_n == 1'b0)
                counter_init <= 32'h0;
        else if ( counter_init < 32'h5f5e100 )
                counter_init <= counter_init + 32'b1;
        else ;
end


// gen a 200kHz CLK for work counter count
always @(posedge sys_clk or negedge sys_rst_n) begin
        if (sys_rst_n == 1'b0)
                clk_200k <= 10'b0;
        else if ((counter_div >= 0  ) && (counter_div < 125))
                clk_200k <= 10'b0;
        else if ((counter_div >= 125) && (counter_div < 250))
                clk_200k <= 10'b1;
        else if ((counter_div >= 250) && (counter_div < 375))
                clk_200k <= 10'b0;
        else if ((counter_div >= 375) && (counter_div < 500))
                clk_200k <= 10'b1;
        else if ((counter_div >= 500) && (counter_div < 625))
                clk_200k <= 10'b0;
```

```verilog
        else if ((counter_div >= 625) && (counter_div < 750))
            clk_200k <= 10'b1;
        else if ((counter_div >= 750) && (counter_div < 875))
            clk_200k <= 10'b0;
        else if ((counter_div >= 875) && (counter_div < 1000))
            clk_200k <= 10'b1;
        else ;
end

// when AT24C16 init finish, work counter start to add
always @(posedge clk_200k or negedge sys_rst_n) begin
        if (sys_rst_n == 1'b0)
            counter <= 8'h0;
        else if ( counter_init == 32'h5f5e100 && counter < 8'hff )
            counter <= counter + 8'b1;
        else ;
end


//generate real clk for SCLK ,when the i2c bus is idle, make the clk wire high level
always @( * ) begin
        if ( counter >= 2 && counter <= 118 )
            i2c_sclk = clk_50k;
        else
            i2c_sclk = 1'b1;
end


// output SDA data with AT24C16 data sheet request
//使用四倍频方便控制 I2C 的时序高低变化,4 个计数代表一个 bit 操作
//下面全部是按照 EEPROM 的时序要求写特定 bit
always @( * ) begin
    case (counter)
        0 : sda = 1'b1 ;
        1 : sda = 1'b1 ;
        2 : sda = 1'b1 ;
        3 : sda = 1'b0 ;
        4 : sda = 1'b0 ;                        //SOP 开始时序
```

```
 5 : sda = 1'b1 ;
 6 : sda = 1'b1 ;
 7 : sda = 1'b1 ;
 8 : sda = 1'b1 ;              // 1

 9 : sda = 1'b0 ;
10 : sda = 1'b0 ;
11 : sda = 1'b0 ;
12 : sda = 1'b0 ;              //0

13 : sda = 1'b1 ;
14 : sda = 1'b1 ;
15 : sda = 1'b1 ;
16 : sda = 1'b1 ;              //1

17 : sda = 1'b0 ;
18 : sda = 1'b0 ;
19 : sda = 1'b0 ;
20 : sda = 1'b0 ;              //0

21 : sda = device_addr[2] ;
22 : sda = device_addr[2] ;
23 : sda = device_addr[2] ;
24 : sda = device_addr[2] ;

25 : sda = device_addr[1] ;
26 : sda = device_addr[1] ;
27 : sda = device_addr[1] ;
28 : sda = device_addr[1] ;

29 : sda = device_addr[0] ;
30 : sda = device_addr[0] ;
31 : sda = device_addr[0] ;
32 : sda = device_addr[0] ;

33 : sda = 1'b0 ;
```

```
34: sda = 1'b0 ;
35: sda = 1'b0 ;
36: sda = 1'b0 ;                    // write ,bit[0] = 0

37: sda = 1'bz ;
38: sda = 1'bz ;
39: sda = 1'bz ;
40: sda = 1'bz ;                    //ACK ,sda should be input

41: sda = memory_addr[7] ;
42: sda = memory_addr[7] ;
43: sda = memory_addr[7] ;
44: sda = memory_addr[7] ;

45: sda = memory_addr[6] ;
46: sda = memory_addr[6] ;
47: sda = memory_addr[6] ;
48: sda = memory_addr[6] ;

49: sda = memory_addr[5] ;
50: sda = memory_addr[5] ;
51: sda = memory_addr[5] ;
52: sda = memory_addr[5] ;

53: sda = memory_addr[4] ;
54: sda = memory_addr[4] ;
55: sda = memory_addr[4] ;
56: sda = memory_addr[4] ;

57: sda = memory_addr[3] ;
58: sda = memory_addr[3] ;
59: sda = memory_addr[3] ;
60: sda = memory_addr[3] ;

61: sda = memory_addr[2] ;
62: sda = memory_addr[2] ;
```

```
63: sda = memory_addr[2];
64: sda = memory_addr[2];

65: sda = memory_addr[1];
67: sda = memory_addr[1];
67: sda = memory_addr[1];
68: sda = memory_addr[1];

69: sda = memory_addr[0];
70: sda = memory_addr[0];
71: sda = memory_addr[0];
72: sda = memory_addr[0];

73: sda = 1'bz;
74: sda = 1'bz;
75: sda = 1'bz;
76: sda = 1'bz;                    //ACK

77: sda = buff[7];
78: sda = buff[7];
79: sda = buff[7];
80: sda = buff[7];                 //wirte data, hign bit 7

81: sda = buff[6];
82: sda = buff[6];
83: sda = buff[6];
84: sda = buff[6];                 //wirte data, hign bit 6

85  sda = buff[5];
86  sda = buff[5];
87: sda = buff[5];
88: sda = buff[5];                 //wirte data, hign bit 5

89: sda = buff[4];
90: sda = buff[4];
91: sda = buff[4];
```

```
       92: sda = buff[4] ;              //wirte data, hign bit 4

       93: sda = buff[3] ;
       94: sda = buff[3] ;
       95: sda = buff[3] ;
       96: sda = buff[3] ;              //wirte data, hign bit 3

       97: sda = buff[2] ;
       98: sda = buff[2] ;
       99: sda = buff[2] ;
      100: sda = buff[2] ;              //wirte data, hign bit 2

      101: sda = buff[1] ;
      102: sda = buff[1] ;
      103: sda = buff[1] ;
      104: sda = buff[1] ;              //wirte data, hign bit 1

      105: sda = buff[0] ;
      106: sda = buff[0] ;
      107: sda = buff[0] ;
      108: sda = buff[0] ;              //wirte data, hign bit 0

      109: sda = 1'bz ;
      110: sda = 1'bz ;
      111: sda = 1'bz ;
      112: sda = 1'bz ;                 //ACK

      113: sda = 1'b0 ;
      114: sda = 1'b0 ;

      115: sda = 1'b1 ;
      116: sda = 1'b1 ;
      117: sda = 1'b1 ;                 //EOP

      default : sda = 1'b1 ;
endcase
```

        end

// output SDA data enable with AT24C16 data sheet request
//使能信号控制 SDA 是否输出
always @(*) begin
    case (counter)
        // 0 : enable = 1'b1 ;   //when the bus is idle,the bus should be release
        1 : enable = 1'b1 ;
        2 : enable = 1'b1 ;
        3 : enable = 1'b1 ;
        4 : enable = 1'b1 ;   //SOP

        5 :enable = 1'b1 ;
        6 :enable = 1'b1 ;
        7 :enable = 1'b1 ;
        8 :enable = 1'b1 ;   // 1

        9 :enable = 1'b1 ;
        10 :enable = 1'b1 ;
        11 :enable = 1'b1 ;
        12 :enable = 1'b1 ;   //0

        13 :enable = 1'b1 ;
        14 :enable = 1'b1 ;
        15 :enable = 1'b1 ;
        16 :enable = 1'b1 ;   //1

        17 :enable = 1'b1 ;
        18 :enable = 1'b1 ;
        19 :enable = 1'b1 ;
        20 :enable = 1'b1 ;   //0

        21 :enable = 1'b1 ;
        22 :enable = 1'b1 ;
        23 :enable = 1'b1 ;
        24 :enable = 1'b1 ;

25 :enable = 1'b1 ;
26 :enable = 1'b1 ;
27 :enable = 1'b1 ;
28 :enable = 1'b1 ;

29 : enable = 1'b1 ;
30 : enable = 1'b1 ;
31 : enable = 1'b1 ;
32 : enable = 1'b1 ;

33 :enable = 1'b1 ;
34 :enable = 1'b1 ;
35 :enable = 1'b1 ;
36 :enable = 1'b1 ;   // write ,bit[0] = 0

41:enable = 1'b1 ;
42:enable = 1'b1 ;
43:enable = 1'b1 ;
44:enable = 1'b1 ;

45:enable = 1'b1 ;
46:enable = 1'b1 ;
47:enable = 1'b1 ;
48:enable = 1'b1 ;

49:enable = 1'b1 ;
50:enable = 1'b1 ;
51:enable = 1'b1 ;
52:enable = 1'b1 ;

53:enable = 1'b1 ;
54:enable = 1'b1 ;
55:enable = 1'b1 ;
56:enable = 1'b1 ;

```
57:enable = 1'b1 ;
58:enable = 1'b1 ;
59:enable = 1'b1 ;
60:enable = 1'b1 ;

61:enable = 1'b1 ;
62:enable = 1'b1 ;
63:enable = 1'b1 ;
64:enable = 1'b1 ;

65:enable = 1'b1 ;
67:enable = 1'b1 ;
67:enable = 1'b1 ;
68:enable = 1'b1 ;

69:enable = 1'b1 ;
70:enable = 1'b1 ;
71:enable = 1'b1 ;
72:enable = 1'b1 ;

77:enable = 1'b1 ;
78:enable = 1'b1 ;
79:enable = 1'b1 ;
80:enable = 1'b1 ;          //wirte data, hign bit 7

81:enable = 1'b1 ;
82:enable = 1'b1 ;
83:enable = 1'b1 ;
84:enable = 1'b1 ;          //wirte data, hign bit 6

85:enable = 1'b1 ;
86:enable = 1'b1 ;
87:enable = 1'b1 ;
88:enable = 1'b1 ;          //wirte data, hign bit 5

89:enable = 1'b1 ;
```

```
            90:enable = 1'b1 ;
            91:enable = 1'b1 ;
            92:enable = 1'b1 ;           //wirte data, hign bit 4

            93:enable = 1'b1 ;
            94:enable = 1'b1 ;
            95:enable = 1'b1 ;
            96:enable = 1'b1 ;           //wirte data, hign bit 3

            97:enable = 1'b1 ;
            98:enable = 1'b1 ;
            99:enable = 1'b1 ;
            100:enable = 1'b1 ;          //wirte data, hign bit 2

            101:enable = 1'b1 ;
            102:enable = 1'b1 ;
            103:enable = 1'b1 ;
            104:enable = 1'b1 ;          //wirte data, hign bit 1

            105:enable = 1'b1 ;
            106:enable = 1'b1 ;
            107:enable = 1'b1 ;
            108:enable = 1'b1 ;          //wirte data, hign bit 0

            113:enable = 1'b1 ;
            114:enable = 1'b1 ;

            115:enable = 1'b1 ;
            116:enable = 1'b1 ;
            117:enable = 1'b1 ;          //EOP
    default : enable = 1'b0 ;
        endcase
end


// output sda data when sda enable is 1, else output Z state
//使用使能信号控制三态接口,应该输出还是释放总线(输出 Z 态)
```

```
assign i2c_sda = (enable == 1'b1)? sda : 1'bz;

// disp 0x55 to the led when write end
always @(posedge sys_clk or negedge sys_rst_n) begin
    if (sys_rst_n == 1'b0)
        LED <= 8'b0;
    else if ( counter == 117 )
        LED <= 8'h55 ;
    else ;
end

endmodule

//end of RTL code
```

3.10.5 实验现象

本实验现象为 EEPROM 写入后,4 个 LED 灯间隔发光,如图 3.43 所示。

图 3.43 EEPROM 写操作实验现象

3.11 EEPROM 读操作实验

3.11.1 EEPROM 读操作简介

EEPROM 读操作还是采用 I2C 接口的 AT24C16 芯片作为示例。

EEPROM 读操作和 EEPROM 写操作差异不大,操作时序基本相同,整体程序框架可以采用 EEPROM 写的程序框架。如果您学会了 EEPROM 写操作的程序,那么 EEPROM 读操作应该可以自己写出来。

3.11.2 实验任务

在 EEPROM AT24C16 的 0x02 地址正确读出上次写入的数据,0x55 数据。

3.11.3 硬件设计

EEPROM 读操作的硬件部分和 EEPROM 写操作部分完全一样,都是需要一个 I2C 接口的 AT24C16 芯片,该芯片在读/写操作时,只看到两个 I/O 口、CL 时钟线和 SDA 双向数据线。在送入读命令的时候,SDA 的方向是从 FPGA 输入到 AT24C16;在读出数据的时候,SDA 的方向是从 AT24C16 输入到 FPGA。

3.11.4 程序设计

1. 设计思路

本次实验也是主要根据 AT24CXX 的字节读时序要求以及 I2C 总线协议的规定来编写程序的。

AT24CXX 的读操作包括立即地址字节读、选择读和连续读。本文采用立即地址字节读模式。

2. 立即地址字节读

AT24CXX 的地址计数器内容为最后操作字节的地址加 1。也就是说,如果上次读/写的操作地址为 N,则立即读的地址从地址 $N+1$ 开始;如果 $N=E$(这里对 24C01,$E=127$;对 24C02,$E=255$;对 24C04,$E=511$;对 24C08,$E=1023$;对 24C16,$E=2047$),则计数器将翻转到 0 且继续输出数据 AT24C01/02/04/08/16 接收到从器件地址信号后 R/W 位置 1。它首先发送一个应答信号,然后发送一个 8 位字数据。主器件不需发送一个应答信号,但要产生一个停止信号字节写时序。

EEPROM 读时序图如图 3.44 所示。

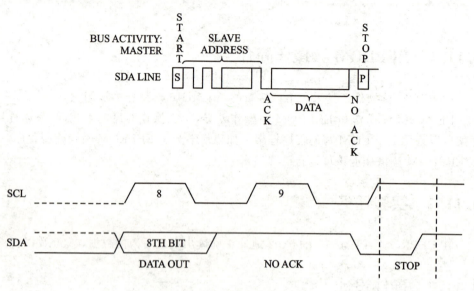

图 3.44　EEPROM 读时序图

时序主要包括起始位、从器件地址、字节地址、数据、ACK 和 STOP 位。读时序和写时序的设计思路一致。

3. 源代码

```
module e2prom_read (

    input               sys_clk,            //system clock
    input               sys_rst_n,          //system reset, low is active;

    inout               i2c_sda,

    //output ports
    output              i2c_sclk,

    output reg [7:0]    LED
);

//AT24C16 device address is 000, this value is no care
assign device_addr = 3'b000;        //device address is 000

// AT24C16 read addr
assign memory_addr = 8'h02;         //memeory address is 02
```

```verilog
// counter for gen a clk_50k ; need count to 1 000, for 50 MHz/1 000 = 50 kHz
always @(posedge sys_clk or negedge sys_rst_n) begin
        if (sys_rst_n == 1'b0)
            counter_div <= 10'b0;
        else if (counter_div >= 10'd999)
            counter_div <= 10'b0;
        else
            counter_div <= counter_div + 10'b1;
end

// gen a clk_50k use counter_div ;not use counter_div 0 - 500 is for i2c bus request
//start timing
always @(posedge sys_clk or negedge sys_rst_n) begin
        if (sys_rst_n == 1'b0)
            clk_50k <= 10'b0;
        else  if ((counter_div >= 375) && (counter_div < 875))
            clk_50k <= 10'b1;
        else
            clk_50k <= 10'b0;
end

// counter for init for AT24C16
always @(posedge sys_clk or negedge sys_rst_n) begin
        if (sys_rst_n == 1'b0)
            counter_init <= 32'h0;
        else if ( counter_init < 32'h5f5e100 )
            counter_init  <= counter_init + 32'b1;
        else ;
end

// gen a 200K CLK for work counter count
always @(posedge sys_clk or negedge sys_rst_n) begin
        if (sys_rst_n == 1'b0)
            clk_200k <= 10'b0;
        else  if ((counter_div >= 0  ) && (counter_div < 125))
            clk_200k <= 10'b0;
        else  if ((counter_div >= 125) && (counter_div < 250))
            clk_200k <= 10'b1;
        else  if ((counter_div >= 250) && (counter_div < 375))
            clk_200k <= 10'b0;
        else  if ((counter_div >= 375) && (counter_div < 500))
            clk_200k <= 10'b1;
```

```verilog
        else  if ((counter_div >= 500) && (counter_div < 625))
            clk_200k <= 10'b0;
        else  if ((counter_div >= 625) && (counter_div < 750))
            clk_200k <= 10'b1;
        else  if ((counter_div >= 750) && (counter_div < 875))
            clk_200k <= 10'b0;
        else  if ((counter_div >= 875) && (counter_div < 1000))
            clk_200k <= 10'b1;
        else ;
    end


// when AT24C16 init finish, work counter start to add
always @(posedge clk_200k or negedge sys_rst_n) begin
        if (sys_rst_n == 1'b0)
            counter <= 8'h0;
        else if ( counter_init == 32'h5f5e100 && counter < 8'hff )
            counter <= counter + 8'b1;
        else ;
    end


//generate real clk for SCLK ,when the i2c bus is idle, make the clk wire high level
//always @( * ) begin
        if ( counter >= 2 && counter <= 156 )
            i2c_sclk = clk_50k;
        else
            i2c_sclk = 1'b1;
    end


// output SDA data with AT24C16 data sheet request

always @( * ) begin
    case (counter)
        0 : sda = 1'b1 ;
        1 : sda = 1'b1 ;
        2 : sda = 1'b1 ;
        3 : sda = 1'b0 ;
        4 : sda = 1'b0 ;                   //SOP

        5 : sda = 1'b1 ;
        6 : sda = 1'b1 ;
        7 : sda = 1'b1 ;
        8 : sda = 1'b1 ;                   // 1
```

9 : sda = 1'b0 ;
10 : sda = 1'b0 ;
11 : sda = 1'b0 ;
12 : sda = 1'b0 ;   //0

13 : sda = 1'b1 ;
14 : sda = 1'b1 ;
15 : sda = 1'b1 ;
16 : sda = 1'b1 ;   //1

17 : sda = 1'b0 ;
18 : sda = 1'b0 ;
19 : sda = 1'b0 ;
20 : sda = 1'b0 ;   //0

21 : sda = device_addr[2] ;
22 : sda = device_addr[2] ;
23 : sda = device_addr[2] ;
24 : sda = device_addr[2] ;

25 : sda = device_addr[1] ;
26 : sda = device_addr[1] ;
27 : sda = device_addr[1] ;
28 : sda = device_addr[1] ;

29 : sda = device_addr[0] ;
30 : sda = device_addr[0] ;
31 : sda = device_addr[0] ;
32 : sda = device_addr[0] ;

33 : sda = 1'b0 ;
34 : sda = 1'b0 ;
35 : sda = 1'b0 ;
36 : sda = 1'b0 ;   //presude write ,bit[0] = 0

37: sda = 1'bz ;
38: sda = 1'bz ;
39: sda = 1'bz ;
40: sda = 1'bz ;   //ACK ,sda should be input

41: sda = memory_addr[7] ;

```
42: sda = memory_addr[7] ;
43: sda = memory_addr[7] ;
44: sda = memory_addr[7] ;

45: sda = memory_addr[6] ;
46: sda = memory_addr[6] ;
47: sda = memory_addr[6] ;
48: sda = memory_addr[6] ;

49: sda = memory_addr[5] ;
50: sda = memory_addr[5] ;
51: sda = memory_addr[5] ;
52: sda = memory_addr[5] ;

53: sda = memory_addr[4] ;
54: sda = memory_addr[4] ;
55: sda = memory_addr[4] ;
56: sda = memory_addr[4] ;

57: sda = memory_addr[3] ;
58: sda = memory_addr[3] ;
59: sda = memory_addr[3] ;
60: sda = memory_addr[3] ;

61: sda = memory_addr[2] ;
62: sda = memory_addr[2] ;
63: sda = memory_addr[2] ;
64: sda = memory_addr[2] ;

65: sda = memory_addr[1] ;
67: sda = memory_addr[1] ;
67: sda = memory_addr[1] ;
68: sda = memory_addr[1] ;

69: sda = memory_addr[0] ;
70: sda = memory_addr[0] ;
71: sda = memory_addr[0] ;
72: sda = memory_addr[0] ;

73: sda = 1'bz ;
74: sda = 1'bz ;
75: sda = 1'bz ;
```

```
76: sda = 1'bz ;                    //ACK

77: sda = 1'b1 ;
78: sda = 1'b1 ;
79: sda = 1'b0 ;
80: sda = 1'b0 ;                    //SOP

81: sda = 1'b1 ;
82: sda = 1'b1 ;
83: sda = 1'b1 ;
84: sda = 1'b1 ;

85: sda = 1'b0 ;
86: sda = 1'b0 ;
87: sda = 1'b0 ;
88: sda = 1'b0 ;

89: sda = 1'b1 ;
90: sda = 1'b1 ;
91: sda = 1'b1 ;
92: sda = 1'b1 ;

93: sda = 1'b0 ;
94: sda = 1'b0 ;
95: sda = 1'b0 ;
96: sda = 1'b0 ;

97: sda = device_addr[2] ;
98: sda = device_addr[2] ;
99: sda = device_addr[2] ;
100: sda = device_addr[2] ;

101: sda = device_addr[1] ;
102: sda = device_addr[1] ;
103: sda = device_addr[1] ;
104: sda = device_addr[1] ;

105: sda = device_addr[0] ;
106: sda = device_addr[0] ;
107: sda = device_addr[0] ;
108: sda = device_addr[0] ;
```

```
109:sda = 1'b1 ;
110:sda = 1'b1 ;
111:sda = 1'b1 ;
112:sda = 1'b1 ;              //Read Enable 1,R/W = 1 时为读数据 方向从-＞主

113:sda = 1'b0 ;
114:sda = 1'b0 ;
115:sda = 1'b0 ;
116:sda = 1'b0 ;              //ACK

117:sda = 1'b0 ;
118:sda = 1'b0 ;
119:sda = 1'b0 ;
120:sda = 1'b0 ;              //DATA[7]

121:sda = 1'b0 ;
122:sda = 1'b0 ;
123:sda = 1'b0 ;
124:sda = 1'b0 ;              //DATA[6]

125:sda = 1'b0 ;
126:sda = 1'b0 ;
127:sda = 1'b0 ;
128:sda = 1'b0 ;              //DATA[5]

129:sda = 1'b0 ;
130:sda = 1'b0 ;
131:sda = 1'b0 ;
132:sda = 1'b0 ;              //DATA[4]

133:sda = 1'b0 ;
134:sda = 1'b0 ;
135:sda = 1'b0 ;
136:sda = 1'b0 ;              //DATA[3]

137:sda = 1'b0 ;
138:sda = 1'b0 ;
139:sda = 1'b0 ;
140:sda = 1'b0 ;              //DATA[2]

141:sda = 1'b0 ;
142:sda = 1'b0 ;
```

```
            143:sda = 1'b0 ;
            144:sda = 1'b0 ;                    //DATA[1]

            145:sda = 1'b1 ;
            146:sda = 1'b1 ;
            147:sda = 1'b1 ;
            148:sda = 1'b1 ;                    //DATA[0]

            149:sda = 1'b0 ;
            150:sda = 1'b0 ;
            151:sda = 1'b0 ;
            152:sda = 1'b0 ;                    //NON ACK

            153:sda = 1'b0 ;
            154:sda = 1'b0 ;

            155:sda = 1'b1 ;
            156:sda = 1'b1 ;
            157:sda = 1'b1 ;                    //EOP
            default:sda = 1'b1 ;
        endcase
end

// output SDA data enable with AT24C16 data sheet request
always @( * ) begin
    case (counter)
        //   0 : sda = 1'b1 ;
             1 : enable = 1'b1 ;
             2 : enable = 1'b1 ;
             3 : enable = 1'b1 ;
             4 : enable = 1'b1 ;                //SOP

             5 : enable = 1'b1 ;
             6 : enable = 1'b1 ;
             7 : enable = 1'b1 ;
             8 : enable = 1'b1 ;                // 1

             9 :enable = 1'b1 ;
            10 :enable = 1'b1 ;
            11 :enable = 1'b1 ;
            12 :enable = 1'b1 ;                 //0
```

```
13 : enable = 1'b1 ;
14 : enable = 1'b1 ;
15 : enable = 1'b1 ;
16 : enable = 1'b1 ;                //1

17 : enable = 1'b1 ;
18 : enable = 1'b1 ;
19 : enable = 1'b1 ;
20 : enable = 1'b1 ;                //0

21 : enable = 1'b1 ;
22 : enable = 1'b1 ;
23 : enable = 1'b1 ;
24 : enable = 1'b1 ;

25 : enable = 1'b1 ;
26 : enable = 1'b1 ;
27 : enable = 1'b1 ;
28 : enable = 1'b1 ;

29 : enable = 1'b1 ;
30 : enable = 1'b1 ;
31 : enable = 1'b1 ;
32 : enable = 1'b1 ;

33 : enable = 1'b1 ;
34 : enable = 1'b1 ;
35 : enable = 1'b1 ;
36 : enable = 1'b1 ;                //presude write ,bit[0] = 0

                                    //37~41,ACK ,sda should be input
41: enable = 1'b1 ;
42: enable = 1'b1 ;
43: enable = 1'b1 ;
44: enable = 1'b1 ;

45: enable = 1'b1 ;
46: enable = 1'b1 ;
47: enable = 1'b1 ;
48: enable = 1'b1 ;

49: enable = 1'b1 ;
```

50: enable = 1'b1 ;
51: enable = 1'b1 ;
52: enable = 1'b1 ;

53: enable = 1'b1 ;
54: enable = 1'b1 ;
55: enable = 1'b1 ;
56: enable = 1'b1 ;

57: enable = 1'b1 ;
58: enable = 1'b1 ;
59: enable = 1'b1 ;
60: enable = 1'b1 ;

61: enable = 1'b1 ;
62: enable = 1'b1 ;
63: enable = 1'b1 ;
64: enable = 1'b1 ;

65: enable = 1'b1 ;
67: enable = 1'b1 ;
67: enable = 1'b1 ;
68: enable = 1'b1 ;

69: enable = 1'b1 ;
70: enable = 1'b1 ;
71: enable = 1'b1 ;
72: enable = 1'b1 ;

// 73～76 为 ACK

77: enable = 1'b1 ;
78: enable = 1'b1 ;
79: enable = 1'b1 ;
80: enable = 1'b1 ;   //SOP

81: enable = 1'b1 ;
82: enable = 1'b1 ;
83: enable = 1'b1 ;
84: enable = 1'b1 ;

85: enable = 1'b1 ;
86: enable = 1'b1 ;
87: enable = 1'b1 ;
88: enable = 1'b1 ;

89: enable = 1'b1 ;

```
             90: enable = 1'b1 ;
             91: enable = 1'b1 ;
             92: enable = 1'b1 ;

             93: enable = 1'b1 ;
             94: enable = 1'b1 ;
             95: enable = 1'b1 ;
             96: enable = 1'b1 ;

             97: enable = 1'b1 ;
             98: enable = 1'b1 ;
             99: enable = 1'b1 ;
            100: enable = 1'b1 ;

            101: enable = 1'b1 ;
            102: enable = 1'b1 ;
            103: enable = 1'b1 ;
            104: enable = 1'b1 ;

            105: enable = 1'b1 ;
            106: enable = 1'b1 ;
            107: enable = 1'b1 ;
            108: enable = 1'b1 ;

            109: enable = 1'b1 ;
            110: enable = 1'b1 ;
            111: enable = 1'b1 ;
            112: enable = 1'b1 ;          //Read Enable   1

            153: enable = 1'b1 ;
            154: enable = 1'b1 ;

            155: enable = 1'b1 ;
            156: enable = 1'b1 ;
            157: enable = 1'b1 ;          //EOP

            default : enable = 1'b0 ;
        endcase
    end

// output sda data when sda enable is 1, else output Z state
assign i2c_sda = (enable == 1'b1)?   sda : 1'bz;

// disp read data to the led when read
//在数据有限中间进行采样,数据最稳定
always @(posedge sys_clk or negedge sys_rst_n) begin
```

```
        if(sys_rst_n == 1'b0)
            LED <= 8'h0;
        else
            case(counter)
                119: LED[7] = i2c_sda;            //DATA[7]
                123: LED[6] = i2c_sda;            //DATA[6]
                127: LED[5] = i2c_sda;            //DATA[5]
                131: LED[4] = i2c_sda;            //DATA[4]
                135: LED[3] = i2c_sda;            //DATA[3]
                139: LED[2] = i2c_sda;            //DATA[2]
                143: LED[1] = i2c_sda;            //DATA[1]
                147: LED[0] = i2c_sda;            //DATA[0]
                default :;
            endcase
end

endmodule

//end of RTL code
```

3.11.5 实验现象

将程序 sof 文件下载到开发板,可以观察到 LED 显示 01010101,将之前写入的 0x55 数据通过 LED 显示出来,如图 3.45 所示。

图 3.45 EEPROM 读操作实验现象

3.12 PS/2 键盘读操作实验

3.12.1 PS/2 接口简介

PS/2 接口(见图 3.46)是在较早计算机上常见的接口之一,用于鼠标、键盘等设备。一般情况下,PS/2 接口的鼠标为绿色,键盘为紫色。当前的计算机一般都不再配有 PS/2 接口,PS/2 键盘也比较少见了。本文之所以介绍 PS/2 接口,主要是因为 PS/2 接口协议简单,适合入门学习。

图 3.46 台式机 PS/2 接口

PS/2 接口是输入装置接口,而不是传输接口。所以 PS2 口根本没有传输速率的概念,只有扫描速率。在 Windows 环境下,PS/2 鼠标的采样率默认为 60 次/秒,USB 鼠标的采样率为 120 次/秒。较高的采样率,理论上可以提高鼠标的移动精度。

PS/2 接口设备不支持热插拔,强行带电插拔有可能烧毁主板。

早期,在 PS/2 键盘中,包含了一个嵌入式的微控制器(如 InDl,8048 系列),以用来执行各项的工作并减少整个系统工作中的负担。微控制器要做的工作就是监测所有的按键,以及当按键被按下或放开时,就回报给主机。

PS/2 有 6 引脚的 mini-DIN 和 5 引脚的 DIN 连接器,每种引脚的定义如表 3.4 和表 3.5 所列。

第 3 章　设计实例

表 3.4　PS/2 5 引脚接口定义

插头	插座	5 引脚 DIN(AT/XT)
		1—时钟
		2—数据
		3—未实现,保留
		4—电源地
		5—电源+5 V

表 3.5　PS/2 6 引脚接口定义

插头	插座	6 引脚 mini-DIN(PS/2)
		1—数据
		2—未实现,保留
		3—电源地
		4—电源+5 V
		5—时钟
		6—未实现,保留

3.12.2　实验任务

PS/2 键盘按键,可以发送键值给 LED 显示(十六进制显示)。

3.12.3　硬件设计

图 3.47 中的 PS/2 接口有 4 个引脚:电源、地、数据和时钟。

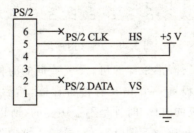

图 3.47　PS/2 键盘读操作实验原理图

PS/2 接口采用一种双向同步串行协议,只需要一个数据口。换句话说,每次数据线上发送一位数据并且每次在时钟线上发一个脉冲就被读入。

本实验是用 PS/2 接口接一个键盘,然后按下键盘按键,通过 PS/2 接口将按键的键值送入 FPGA 逻辑,进行采样处理。

3.12.4 程序设计

1. 设计思路

数据线和时钟线都是集电极开路结构(正常情况为高电平)。当键盘或鼠标等待发送数据时,它首先检查时钟以确认它是否是高电平。如果不是,那么是主机抑制了通信,设备必须缓冲任何要发送的数据,直到重新获得总线的控制权(键盘有 16 字节的缓冲区,而鼠标的缓冲区仅存储最后一个要发送的数据包)。如果时钟线是高电平,设备就可以开始传送数据。

设备到主机通信过程中,键盘和鼠标使用一种包含 11 位的串行协议。即所有数据安排在字节中,每个字节为一帧,包含 11 位。这些位的含义如表 3.6 所列。

表 3.6 PS/2 时序定义

1 star bit. This is always 0.	1 个起始位,总是为 0
8 data bits, least significant bit first.	8 个数据位,低位在前
1 parity bit(odd parity)	1 个校验位,奇校验
1 stop bit. This is always1.	1 个停止位,总是为 1
1 acknowledge bit(Host-to-device cpmmunication only)	1 个应答位(仅在主机对设备的通信中)

每位在时钟下降沿被读入,如图 3.48 所示。
表 3.6 是图 3.51 的各个数据域的解析。

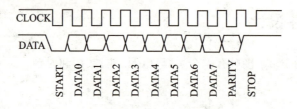

图 3.48 PS/2 时序图

我们只需要在时钟的下降沿采样即可,没有数据的时候,时钟为高电平,所以可以用时钟沿来控制采样。设置一个 buff 移位寄存器,然后在每个时钟下降沿采样,不需要控制信号,没有按键的时候,时钟是不会跳变的,所以只需从 buff 中按顺序取出我们需要的键值即可。每按一下键,就会发送一个键值给上位机。不按键不会发送任何数据。

2. 源代码

module ps2key(

```verilog
                        //input
input           sys_clk,        //system clock
input           sys_rst_n,      //system reset, low is active
                        
input           ps2_key_clk,    //ps2 beyboard clock
input           ps2_key_data,   //ps2 beyboard data
                        //output
output reg [7:0]    led     // output led,in hex format
                );

//reg define
reg     [10:0] buff;        // for regsiger data

//wire define

/***********************************************
**                    Main Program
***********************************************/
//在时钟的下降沿采样,使用移位寄存器串转并即可
always @(negedge ps2_key_clk ) begin
    buff[0]<= ps2_key_data;
    buff[1]<= buff[0];
    buff[2]<= buff[1];
    buff[3]<= buff[2];
    buff[4]<= buff[3];
    buff[5]<= buff[4];
    buff[6]<= buff[5];
    buff[7]<= buff[6];
    buff[8]<= buff[7];
    buff[9]<= buff[8];
    buff[10]<= buff[9];
end

//取有效数据
always @(*) begin
    data_out[7]<= buff[2];
    data_out[6]<= buff[3];
    data_out[5]<= buff[4];
    data_out[4]<= buff[5];
    data_out[3]<= buff[6];
    data_out[2]<= buff[7];
    data_out[1]<= buff[8];
```

```verilog
        data_out[0]<= buff[9];
    end

//assign led = data_out ;              //led is high acitve

always @(*) begin
    led <= data_out ;
end

endmodule
//end of RTL code
```

/*************键盘上各个键的码表,十六进制格式**************/

```
//    0x1C,    <=======>    'a',
//    0x32,    <=======>    'b',
//    0x21,    <=======>    'c',
//    0x23,    <=======>    'd',
//    0x24,    <=======>    'e',
//    0x2B,    <=======>    'f',
//    0x34,    <=======>    'g',
//    0x33,    <=======>    'h',
//    0x43,    <=======>    'i',
//    0x3B,    <=======>    'j',
//    0x42,    <=======>    'k',
//    0x4B,    <=======>    'l',
//    0x3A,    <=======>    'm',
//    0x31,    <=======>    'n',
//    0x44,    <=======>    'o',
//    0x4D,    <=======>    'p',
//    0x15,    <=======>    'q',
//    0x2D,    <=======>    'r',
//    0x1B,    <=======>    's',
//    0x2C,    <=======>    't',
//    0x3C,    <=======>    'u',
//    0x2A,    <=======>    'v',
//    0x1D,    <=======>    'w',
//    0x22,    <=======>    'x',
//    0x35,    <=======>    'y',
//    0x1A,    <=======>    'z',
//    0x45,    <=======>    '0',
//    0x16,    <=======>    '1',
```

```
//    0x1E,    < = = = = = = >        '2',
//    0x26,    < = = = = = = >        '3',
//    0x25,    < = = = = = = >        '4',
//    0x2E,    < = = = = = = >        '5',
//    0x36,    < = = = = = = >        '6',
//    0x3D,    < = = = = = = >        '7',
//    0x3E,    < = = = = = = >        '8',
//    0x46,    < = = = = = = >        '9',
//    0x0E,    < = = = = = = >        ' ',
//    0x4E,    < = = = = = = >        '-',
//    0x55,    < = = = = = = >        '=',
//    0x5D,    < = = = = = = >        '\\',
//    0x29,    < = = = = = = >        ' ',
//    0x54,    < = = = = = = >        '[',
//    0x5B,    < = = = = = = >        ']',
//    0x4C,    < = = = = = = >        ';',
//    0x52,    < = = = = = = >        '\'',
//    0x41,    < = = = = = = >        ',',
//    0x49,    < = = = = = = >        '.',
//    0x4A,    < = = = = = = >        '/',
//    0x71,    < = = = = = = >        '.',
//    0x70,    < = = = = = = >        '0',
//    0x69,    < = = = = = = >        '1',
//    0x72,    < = = = = = = >        '2',
//    0x7A,    < = = = = = = >        '3',
//    0x6B,    < = = = = = = >        '4',
//    0x73,    < = = = = = = >        '5',
//    0x74,    < = = = = = = >        '6',
//    0x6C,    < = = = = = = >        '7',
//    0x75,    < = = = = = = >        '8',
//    0x7D,    < = = = = = = >        '9',
//    0x0d,    < = = = = = = >        ' ',

/***************键盘上各个键的码表,结束***************/
```

3.12.5 实验现象

将 PS/2 键盘插到开发板上,上电,下载 sof 文件到板子上,然后按 Q 键,采样到的真实数据如图 3.49 所示。

data_out 采样稳定后显示为十六进制的 15h,这个是按 Q 键产生的一个键值,和上面键值表比较知道采样是正确的。

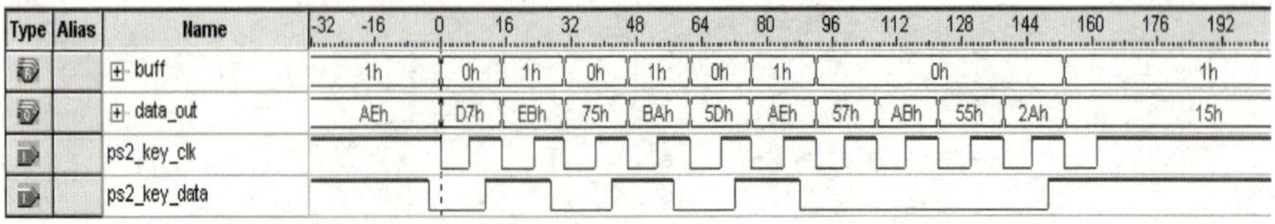

图 3.49　PS/2 键盘读操作实验实现波形图

3.13　VGA 实验

3.13.1　VGA 简介

VGA 是 IBM 公司于 1987 年提出的一个使用模拟信号的计算机显示标准。VGA 接口即计算机采用 VGA 标准输出数据的专用接口。VGA 接口共有 15 针，分成 3 排，每排 5 个孔，是显示器上应用最为广泛的接口类型，绝大多数显示器和投影仪都带有此种接口。

图 3.50 是 VGA 接口实物图。

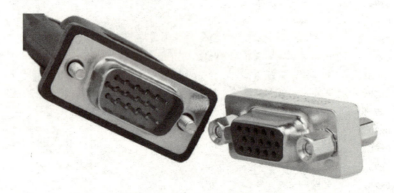

图 3.50　VGA 接口实物图

图 3.51 是 VGA 母座的接口定义。

VGA 接口传输红、绿、蓝模拟信号以及同步信号（水平和垂直信号）。

标准 VGA 一共 15 个接口，真正用到的信号就 5 个。HS（HSYNC）是行同步信号，VS（VSYNC）是场同步信号，VGA_R、VGA_G、VGA_B 是三原色信号，这三个信号接口都是输入模拟信号。我们的 FPGA 开发板直接用 I/O 口去连接 VGA 的 5 个信号接口，并且三原色信号接口输入的只可能是数字信号（0 或 1），因此驱动液晶显示器上显示的颜色最多也就 8 种。

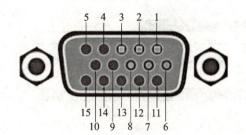

1—红基色；2—绿基色；3—蓝基色；
5—自测试；6—红地；7—绿地；
8—蓝地；9—电源；10—数字地；
13—行同步；14—场同步；
4,11,12,15—地址码

图 3.51 VGA 母座的接口定义

如果想显示更多的颜色,可以在 FPGA 和 VGA 接口间加一个 DA 芯片(比如 ADV7123)的设计,这样就可以实现 65 536 种或者更多种配色效果。

1. VGA 驱动原理

显示从左到右(受水平同步信号 HSYNC 控制)、从上到下(受垂直同步信号 VSYNC 控制)做有规律的移动。屏幕从左上角一点开始,从左到右逐点扫描(显示),每扫描完一行,又重新回到屏幕左边下一行起始位置开始扫描。扫描完所有行,形成一帧时,用场同步信号进行场同步,扫描又回到屏幕左上方,如图 3.52 所示。

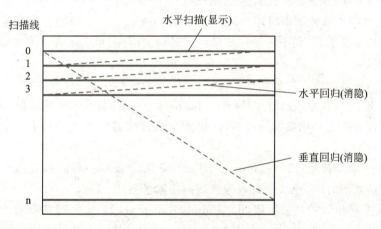

图 3.52 VGA 扫描示意图

完成一行扫描所需的时间称为水平扫描时间,其倒数称为行频率;完成一帧(整屏)扫描所需时间称为垂直扫描时间,其倒数为垂直扫描频率,又称为刷新频率,即刷新一屏的频率。常见的有 60 Hz、75 Hz 等。

VGA 时序图如图 3.53 所示。

上半部分是一个总体的时序图,由 VS 和 HS 构成,下半部分为总体时序图的细化描述。

显示有两个主要时间脉冲:垂直同步脉冲和水平同步脉冲,分别用于控制帧显示与行显示。

图 3.53 VGA 时序图

垂直同步脉冲由三部分组成：

① 垂直同步脉冲开始时序(vertical back porch)表示垂直同步脉冲开始到一帧的有效像素数据开始前的一段时序，也表示有效像素数据开始时不显示的行数。

② 垂直同步脉冲帧时序(vertical active line)表示一帧的有效像素数据开始前到一帧结束的时序，也表示有效像素数据行数。

③ 垂直同步脉冲结束时序(vertical front porch)表示一帧的有效像素数据从开始显示到结束显示后，到下一帧同步脉冲开始前的时序，也表示有效像素数据结束后不显示的行数。

水平同步脉冲由三部分组成：

① 水平同步脉冲开始时序(horizontal back porch)表示水平同步脉冲开始到一行的有效像素数据开始前的一段时序，也表示有效像素数据开始时不显示的像素个数。

② 水平同步脉冲行时序(horizontal active line)表示一行的有效像素数据开始前到一行结束的时序，也表示有效像素数据像素个数。

③ 水平同步脉冲结束时序(horizontal front porch)表示一行的有效像素数据开始结束后到下一行同步脉冲开始前的时序，也表示有效像素数据结束后不显示的像素个数。

2. VGA 显示协议标准

以 640×480 为例，图 3.54 中详细列出了 VGA 时序的各个参数值。

Vertical Region	Scanlines
Front Porch	10
VSync Width	2
Back Porch	33
Display Area	480
Frame Height	525

Horizontal Region	Pixels
Front Porch	16
HSync Width	96
Back Porch	48
Display Area	640
Scanline Width	800

Symbol	Parameter	Vertical Sync			Horizontal Sync	
		Time	Clocks	Lines	Time	Clocks
T_S	Sync pules time	16.7 ms	416 800	521	32 μs	800
T_{DISP}	Display time	15.36 ms	348 000	480	25.6 μs	640
T_{PW}	Pulse width	64 μs	1 600	2	3.84 μs	96
T_{FP}	Front porch	320 μs	8 000	10	640 ns	16
T_{BP}	Back porch	928 μs	23 200	29	1.92 μs	48

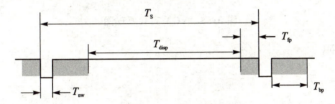

图 3.54　VGA 显示协议标准

3.13.2　实验任务

驱动 VGA 液晶显示器显示固定颜色。

3.13.3　硬件设计

本实验的 VGA 显示需要关注 VGA 原理图里面的 5 个信号,分别是 HS/VS 同步信号,R、G、B 颜色信号。本书采用的开发板使用 3 bit FPGA I/O 作为 RGB 信号线,可以产生 8 个颜色。图 3.55 所示为 VGA 实验。

3.13.4　程序设计

1. 设计思路

行显示→行消隐前肩→行消隐→行消隐后肩。
场显示→场消隐前肩→场消隐→场消隐后肩。

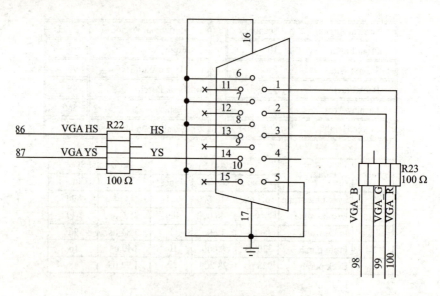

图 3.55 VGA 实验原理图

2. 源代码

```
module vga_simple
(
    input     sys_clk,           //系统时钟
    input     sys_rst_n,         //异步复位信号

    output reg [2:0]   VGA_RGB,  //接收要显示的色彩
    output wire        VGA_HS,   // VGA 引脚 行同步
    output wire        VGA_VS    // VGA 引脚 场同步

);

reg [9:0] h_cnt;
reg [9:0] v_cnt;
reg VGA_CLK;

parameter   h_a = 96;
parameter   h_b = 40;
parameter   h_c = 640;
parameter   h_d = 24;

parameter   v_o = 2;
parameter   v_p = 33 ;
```

```verilog
parameter    v_q = 480;
parameter    v_s = 10 ;

always @ (posedge sys_clk or negedge sys_rst_n) begin
    if (! sys_rst_n)
        VGA_CLK <= 1'b0;
    else
        VGA_CLK <= ~VGA_CLK;
end

//行同步信号发生器
always @ (posedge VGA_CLK or negedge sys_rst_n) begin
    if(sys_rst_n == 1'b0)
        h_cnt <= 10'b0;
    else begin
        if(h_cnt >= h_a + h_b + h_c + h_d - 1)
            h_cnt <= 10'b0;
        else
            h_cnt <= h_cnt + 1;
    end
end

//场同步信号发生器
always @ (posedge VGA_CLK or negedge sys_rst_n) begin
    if(sys_rst_n == 1'b0)
        v_cnt <= 10'b0;
    else begin
        if(h_cnt == 10'b0) begin
            if(v_cnt >= v_o + v_p + v_q + v_s - 1)
                v_cnt <= 10'b0;
            else
                v_cnt <= v_cnt + 1;
        end
    end
end

assign VGA_VS = (v_cnt >= 0 && v_cnt < v_o) ? 1'b0 : 1'b1;
assign VGA_HS = (h_cnt >= 0 && h_cnt < h_a) ? 1'b0 : 1'b1;

//显示红色
always @ (posedge VGA_CLK or negedge sys_rst_n) begin
    if(sys_rst_n == 1'b0)
```

```
                VGA_RGB <= 3'b000;
        else if (v_cnt >= v_o+v_p && v_cnt < v_o+v_p+480) begin
            if(h_cnt >= h_a+h_b && h_cnt < h_a+h_b+640)
                VGA_RGB <= 3'b100 ;
            else
                VGA_RGB <= 3'b000;
        end
        else
            VGA_RGB <= 3'b000;
    end

endmodule
```

3.13.5 实验现象

程序下载到 FPGA 开发板上,显示器固定显示红色。图 3.56 所示为 VGA 实验现象。

图 3.56 VGA 实验现象(实际显示器显示红色)

3.14 LCD1602 实验

3.14.1 LCD1602 简介

字符型液晶显示模块是一种专门用于显示字母、数字、符号等点阵式 LCD,目前常用 16×1,16×2,20×2 和 40×2 行相等的模块,其中 16×2 指的是 LCD1602,这种液晶用的非常普遍。

LCD1602 字符型液晶显示器实物如图 3.57 所示。

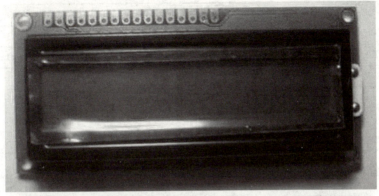

图 3.57　LCD1602 实物图

LCD1602 字符型液晶参数如表 3.7 和表 3.8 所列(摘自长沙太阳人电子有限公司 SMC1602A LCM 手册)。

表 3.7　主要技术参数

显示容量	16×2 个字符
芯片工作电压/V	4.5～5.5
工作电流/mA	2.0(5.0 V)
模块最佳工作电压/V	5.0
字符尺寸 W×H	2.95 mm×4.35 mm

表 3.8　接口信号说明

编号	符号	引脚说明	编号	符号	引脚说明
1	VSS	电源地	9	D2	Data I/O
2	VDD	电源正极	10	D3	Data I/O
3	VL	液晶显示偏压信号	11	D4	Data I/O
4	RS	数据/命令选择端(H/L)	12	D5	Data I/O
5	R/W	读/写选择端(H/L)	13	D6	Data I/O
6	E	使能信号	14	D7	Data I/O
7	D0	Data I/O	15	BLA	背光源正极
8	D1	Data I/O	16	BLK	背光源负极

本实验对 LCD1602 只涉及写操作,操作时序如图 3.58 所示,时序参数如表 3.9 所列。

写操作时序需要在 RW 指示为写(0 代表写),E 使能信号为 1 时,写数据需要保持稳定。

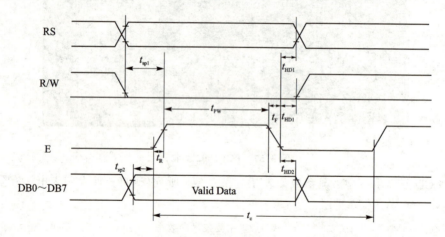

图 3.58　写操作时序

表 3.9　时序参数

时序参数	符号	极限值			单位	测试条件
		最小值	典型值	最大值		
E 信号周期	t_c	400	—	—	ns	引脚 E
E 脉冲宽度	t_{pw}	150	—	—	ns	
E 上升沿/下降沿时间	t_R, t_F	—	—	25	ns	

续表 3.9

时序参数	符 号	极限值			单 位	测试条件
		最小值	典型值	最大值		
地址建立时间	t_{sp1}	30	—	—	ns	引脚 E、RS、R/W
地址保持时间	t_{HD1}	10	—	—	ns	
数据建立时间(读操作)	t_D	—	—	100	ns	引脚 DB0～DB7
数据保持时间(读操作)	t_{HD2}	20	—	—	ns	
数据建立时间(写操作)	t_{SP2}	40	—	—	ns	
数据保持时间(写操作)	t_{HD2}	10	—	—	ns	

3.14.2　实验任务

在 LCD1602 液晶模块上显示字符串,其中第一行显示"Welcome to LCD",第二行显示" by DongDong"。

3.14.3　硬件设计

图 3.59 所示为 LCD1602 实验原理图。

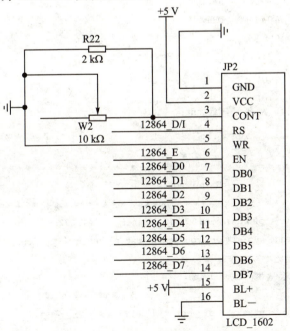

图 3.59　LCD1602 实验原理图

JP2 为 LCD1602 的插座，LCD1602 引脚和 LCD12864 复用部分引脚，然后直接连到 FPGA I/O。

3.14.4 程序设计

1. 设计思路

LCD1602 的设计思路比较清晰，就是按照 LCD1602 的时序要求操作即可，先进行初始化，然后进行写 LCD1602 内部 RAM 显示字符，每个操作，代码中都有详细注释，在此不再详述。

2. 源代码

```verilog
module LCD1602_TOP (
// input
input       sys_clk,
input       sys_rst_n,

// output
output  wire        LCD_EN,
output  wire        RS,
output  wire        RW,
output  wire [7:0]  DB8
);

// DEFINE

reg     [16:0]      div_cnt;
reg                 clk_lcd;

/**************************************************
 *                  Main Program
 **************************************************/
// 50M/10_0000 = 500 Hz, need 17 bit Counter == 131 072
always @(posedge sys_clk or negedge sys_rst_n) begin
        if (sys_rst_n == 1'b0)
            div_cnt <=  17'b0;
        else
            div_cnt <=  div_cnt + 17'b1;
end
```

```verilog
// LCD 时钟频率在 500 Hz 左右即可
always @(posedge sys_clk or negedge sys_rst_n) begin
        if (sys_rst_n == 1'b0)
            clk_lcd <=    1'b0;
        else if ( div_cnt == {17{1'b1}} )
            clk_lcd <=    ~clk_lcd;
        else ;
end

LCD_Driver   LCD_Driver_U0 (
            .clk_lcd   (clk_lcd),
            .rst       (sys_rst_n),
            .LCD_EN    (LCD_EN),
            .RS        (RS),
            .RW        (RW),
            .DB8       (DB8)
            );

endmodule

//功能简述:在 1602 液晶模块上显示字符串,其中第一行显示"Welcome to LCD",
//第二行显示" by DongDong"
//液晶模块为 1602A,相关特性请参考 LCD1602A 数据手册

module LCD_Driver (
// input
input   clk_lcd,
input   rst,

// output
output   wire          LCD_EN,
output   reg           RS,
output   wire          RW,
output   reg [7:0]     DB8
);

// LCD_EN 为 LCD 模块的使能信号(下降沿触发)
// RS = 0 时为写指令;RS = 1 时为写数据
// RW = 0 时对 LCD 模块执行写操作;
// RW = 1 时对 LCD 模块执行读操作

reg     [111:0] Data_First_Buf;
```

```verilog
reg     [111:0] Data_Second_Buf;
//液晶显示的数据缓存

reg     LCD_EN_Sel;
reg     [3:0] disp_count;
reg     [3:0] state;

parameter   Clear_Lcd = 4'b0000;
//清屏并光标复位

parameter   Set_Disp_Mode = 4'b0001;
//设置显示模式:8 位 2 行 5×7 点阵

parameter   Disp_On = 4'b0010;
//显示器开、光标不显示、光标不允许闪烁

parameter   Shift_Down = 4'b0011;
//文字不动,光标自动右移

parameter   Write_Addr = 4'b0100;
//写入显示起始地址

parameter   Write_Data_First = 4'b0101;
//写入第一行显示的数据

parameter   Write_Data_Second = 4'b0110;
//写入第二行显示的数据

parameter   Idel = 4'b0111;
//空闲状态

parameter   Data_First = "Welcome to LCD";
//液晶显示第一行的数据

parameter   Data_Second = "by DongDong";
//液晶显示第二行的数据

assign  RW = 1'b0;
//RW = 0 时对 LCD 模块执行写操作

assign  LCD_EN = LCD_EN_Sel ? clk_lcd : 1'b0;
//通过 LCD_EN_Sel 信号来控制 LCD_EN 的开启与关闭
```

```verilog
always @(posedge clk_lcd or negedge rst)
begin
    if(! rst)
        begin
            state <= Clear_Lcd;
            //复位:清屏并光标复位
            RS <= 1'b0;
            //复位:RS=0 时为写指令
            DB8 <= 8'b0;
            //复位:使 DB8 总线输出全 0
            LCD_EN_Sel <= 1'b1;
            //复位:开启液晶使能信号

            disp_count <= 4'b0;
        end
    else
        case(state)
        //初始化 LCD 模块
        Clear_Lcd: begin
                state <= Set_Disp_Mode;
                DB8 <= 8'b00000001;
    //清屏并光标复位
        end
        Set_Disp_Mode: begin
                state <= Disp_On;
                DB8 <= 8'b00111000;
                //设置显示模式:8 位 2 行 5×8 点阵
        end
        Disp_On: begin
                state <= Shift_Down;
                DB8 <= 8'b00001100;
                //显示器开、光标不显示、光标不允许闪烁
        end
        Shift_Down: begin
                state <= Write_Addr;
                DB8 <= 8'b00000110;
    //文字不动,光标自动右移
        end
        Write_Addr: begin
                state <= Write_Data_First;
                DB8 <= 8'b10000001;
```

```verilog
                    //写入第一行显示起始地址:第一行第二个位置
                    Data_First_Buf <= Data_First;
    //将第一行显示的数据赋给 Data_First_Buf?
    end
Write_Data_First: begin
    //写第一行数据
            if(disp_count == 14) begin
            //disp_count 等于 14 时表示第一行数据已写完
                    DB8 <= 8'b11000001;
                    //送入写第二行的指令
                    RS <= 1'b0;
                    disp_count <= 4'b0;
                    Data_Second_Buf <= Data_Second;
                    state <= Write_Data_Second;
                    //写完第一行进入写第二行状态
            end
            else begin
                    DB8 <= Data_First_Buf[111:104];
                    Data_First_Buf <= (Data_First_Buf << 8);
                    RS <= 1'b1;              //RS=1 表示写数据
                    disp_count <= disp_count + 1'b1;
                    state <= Write_Data_First;
            end
    end
Write_Data_Second: begin
    //写第二行数据
            if(disp_count == 14) begin
                    LCD_EN_Sel <= 1'b0;
                    RS <= 1'b0;
                    disp_count <= 4'b0;
                    state <= Idel;           //写完进入空闲状态
            end
            else begin
                    DB8 <= Data_Second_Buf[111:104];
                    Data_Second_Buf <= (Data_Second_Buf << 8);
                    RS <= 1'b1;
                    disp_count <= disp_count + 1'b1;
                    state <= Write_Data_Second;
            end
    end
Idel:
            state <= Idel;           //在 Idel 状态循环
```

```
            default: state <= Clear_Lcd;  //若 state 为其他值,则将 state 置为 Clear_Lcd
        endcase
    end
endmodule
```

3.14.5 实验现象

将 LCD1602 插到开发板上,上电,Quartus 全编译完程序后,下载 sof 文件到板子上,程序运行正确,则 LCD1602 会如图 3.60 显示。

图 3.60 LCD1602 实验现象

3.15 红外遥控实验

3.15.1 红外遥控简介

本书采用的开发板使用的红外接收头为 HS0038B,其实物图如图 3.61 所示。红外遥控器如图 3.62 所示,右图为对应的按键编码。

红外遥控器发射的信号由一串 0 和 1 的二进制代码组成。不同的红外芯片对 0 和 1 的编码有所不同。通常有曼彻斯特编码和脉冲宽度编码。HS0038B 的 0 和 1 采用 PWM 方法编码,即脉冲宽度调制,0 码由 0.56 ms 低电平和 0.565 ms 高电平组合而成,脉冲宽度为 1.125 ms。1 码由 0.56 ms 低电平和 1.69 ms 高电平组合而成。脉冲宽度为 2.25 ms。在编写解码程序时,通过判断脉冲的宽度,即可得到 0 或 1。红外摇控时序格式参考图 3.63。

当按下遥控器的按键时,遥控器将发出如图 3.63 所示的一串二进制代码,称为

图 3.61 HS0038B 实物图

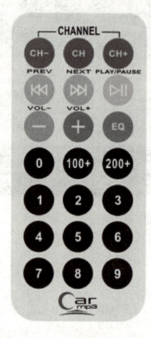

图 3.62 红外遥控器按键编码

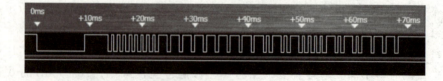

图 3.63 红外遥控时序图

一帧数据。根据各部分的功能,可将它们分为 5 部分,分别为引导码、用户码(8 位)、用户反码(8 位)、数据码(8 位)、数据反码(8 位),编码总共 32 位。遥控器发射代码时,均是低位在前,高位在后。由图 3.63 分析可以得到,引导码低电平为 9 ms,高电平为 4.5 ms,当接收到此码时,表示一帧数据的开始,FPGA 可以准备接收下面的数据。地址码由 8 位二进制组成,共 256 种,图 3.63 中用户反码主要是加强遥控器的可靠性,不同的设备可以拥有不同的地址码。因此,同种编码的遥控器只要设置地址码不同,也不会相互干扰。在同一个遥控器中,所有按键发出的地址码都是相同的。数据码为 8 位,可编码有 256 种状态,代表实际所按下的键。数据反码是数据码的各位求反,通过比较数据码与数据反码,可判断接收到的数据是否正确。如果数据码与数据反码之间的关系不满足相反的关系,则表示本次遥控接收有误,数据应丢弃。在同一个遥控器上,所有按键的数据码均不相同。

遥控器发送的一帧数据共 32 位,其中数据码是 16~31 位,提取这 8 位数据码给 8 个 LED 灯,当按下遥控器的不同按键时,看到不同的 LED 灯亮灭,说明数据接收成功。

以下是具体的实现方法:

开发板上用的是 50 MHz 系统时钟,一个时钟周期是 20 ns,首先通过分频得到 0.125 ms 周期的时钟(div_clk),然后在这个时钟下进行红外信号的接收采样,再定义一个分频计数器(div_cnt),div_clk 使用 div_cnt 产生。

从图 3.63 可以看到,红外一次传输分为 5 部分,分别为引导码、用户码(8 位)、用户反码(8 位)、数据码(8 位)、数据反码(8 位),可以使用一个状态机来完成设计。

状态机根据红外一次传输可以定义五个状态:

- IDLE:空闲状态;
- CHECK_START_9MS:检验 9 ms;
- CHECK_START_4MS:检验 4 ms;
- CHECK_USER_CODE:检验用户码;
- CHECK_DATA_CODE:检验数据码。

首先,当红外信号跳变为低电平时开始计数,直到信号变为高电平,判断这段时间低电平持续时间,如果是 9 ms,则开始下一状态,否则继续执行此状态;在检验 4.5 ms 状态时,信号为高电平时重新开始计数,信号变为低电平时结束计数,判断这段时间的长短,如果满足 4.5 ms 则进入下一个状态——CHECK_USER_CODE 状态。

检验用户码状态时,使用一个 user_cnt 计数器,检测红外信号上升沿的个数,达到 16 个就说明用户码接收完成,可以进入下一个状态数据码的接收。

检验数据码状态时,使用一个 data_cnt 计数器,检测红外信号上升沿的个数,达到 16 个就说明数据码接收完成。数码管接收需要判断电平长短,因为只有知道数据码是 0 还是 1,才能获得数据,所以还得设置一个计数器——data_judge_cnt,用于计数高电平时间,判断数据是 0 还是 1。

最后使用一个移位的寄存器(一般串行数据转换为并行数据都使用移位寄存器),将红外数据依次移入寄存器存储起来,在数据码开始的时候开始移位,根据 data_judge_cnt 的长短判读数据位是 0 还是 1。

最后在移位寄存器接收完成时,使用 LED 锁存有效的数据,此时 LED 灯就得到有效的数据信号了,后续拿来做其他的驱动(比如红外控制数码管、红外控制 VGA 等)都是可以的。

需要说明的是,按键的时候,红外的传输信号有时存在不规范的地方,有非法的数据信号产生。下面是使用 SignalTap 抓波形看到的数据,可以看出,正常的数据传输完后,后面过了一段时间出现了一次非法的数据,如图 3.64 中最后出现的一些的杂波信号。

图 3.64　红外遥控干扰数据(1)

所以我们使用了一个 noise_cnt,用来规避非法的数据信号,也就是每次数据传输完后,我们都要等一段时间,然后才开始进入正常的处理。

下面是完整的波形,如图 3.65 所示。

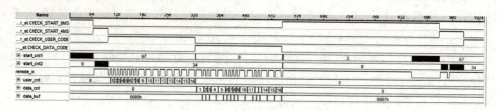

图 3.65　红外遥控干扰数据(2)

当按下遥控器的按键时,如果可以看到不同的按键点亮不同的 LED,则表示实验成功!

3.15.2　实验任务

按下红外遥控器的"一"时,会有三个灯在亮,和红外遥控器简介里面的按键编码相同(三个灯代表十六进制的按键编码 7)。

按下遥控器的不同按键时可以看到不同的 LED 灯亮灭,同一个按键按下时 LED 灯发光的个数是一致的;否则如果每次按同样的按键,LED 灯发光个数不同,那么说明程序、硬件等可能是错误的。

3.15.3　硬件设计

红外遥控实验电路非常简单,只有一个电源、地和数据信号,如图 3.66 所示。当按下红外遥控器按键,HS0038 接收到红外信号后,会通过 DATA 信号送出一个红外遥控的键值。该键值是以固定红外编码的形式送出。FPGA 逻辑需要对该编码时序进行采样解析。

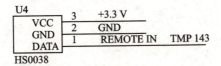

图 3.66　红外遥控实验电路原理图

3.15.4　程序设计

```
module remote_rcv (

//input
input    sys_clk,              //系统时钟
input    sys_rst_n,            //系统复位信号,低电平有效
input    remote_in,            //红外接收信号

//output

output reg [15:0] data_buf,    // for not optimize
output reg [ 7:0] led          //指示 LED
                );

//reg define

reg [11:0]    div_cnt;         //分频计数器
reg           div_clk;
reg           remote_in_dly;
reg [6:0]     start_cnt;
reg [6:0]     start_cnt1;
reg [5:0]     start_cnt2;
reg [5:0]     user_cnt;
reg [5:0]     data_cnt;
reg [14:0]    data_judge_cnt;

reg [14:0]    noise_cnt;

reg [4:0]     curr_st;
reg [4:0]     next_st;


//wire define
```

```verilog
    wire        remote_pos;
    wire        remote_neg;

    //fsm define
    parameter   IDLE = 3'b000;
    parameter   CHECK_START_9MS = 3'b001;
    parameter   CHECK_START_4MS = 3'b010;
    parameter   CHECK_USER_CODE = 3'b011;
    parameter   CHECK_DATA_CODE = 3'b100;


/***************************************************************
 **                       Main Program
 ***************************************************************/

// div cnt  for gen div_clk, 50 MHz/(2×3125) = 0.125 Ms
// 1 div_clk CYCLE need 2 timse div_cnt = 3125
    always @(posedge sys_clk or negedge sys_rst_n) begin        //0.125 ms
        if ( sys_rst_n == 1'b0 )
            div_cnt <= 12'd0;
        else if( div_cnt == 12'd3125 )
            div_cnt <= 12'd0;
        else
            div_cnt = div_cnt + 12'b1;
    end


// 1 div_clk CYCLE need 2 timse div_cnt = 3125
    always @( posedge sys_clk or negedge sys_rst_n) begin
        if ( sys_rst_n == 1'b0 )
            div_clk <= 1'b0;
        else if ( div_cnt == 12'd3125 )
            div_clk <= ~div_clk;
        else ;
    end


//generate FSM next state
    always @(posedge div_clk or negedge sys_rst_n ) begin
            if ( sys_rst_n == 1'b0)
                curr_st <= 1'b0;
```

```verilog
            else
                curr_st <= next_st;
end

//FSM state logic
always @(*) begin
    case (curr_st)
        IDLE: begin
                if ( remote_in == 1'b0 && noise_cnt == 15'd800 )  // noise_cnt is for
                    //clear Remoto signal unvld noise
                        next_st = CHECK_START_9MS;
                    else
                        next_st = IDLE;
        end
        CHECK_START_9MS: begin
                if ( start_cnt1 > 65)
                        next_st = CHECK_START_4MS;
                    else
                        next_st = CHECK_START_9MS;
        end
        CHECK_START_4MS: begin
                if ( start_cnt2 >= 33 )
                        next_st = CHECK_USER_CODE;
                    else
                        next_st = CHECK_START_4MS;
        end
        CHECK_USER_CODE: begin
                if ( user_cnt >= 16 && remote_in == 1'b0 )
                        next_st = CHECK_DATA_CODE;
                    else
                        next_st = CHECK_USER_CODE;
        end
        CHECK_DATA_CODE: begin
            if ( data_cnt >= 16 && remote_in == 1'b0 )
                        next_st = IDLE;
                    else
                        next_st = CHECK_DATA_CODE ;
        end
        default: next_st = IDLE;
```

```verilog
            endcase
    end
    
    
    // for clear Remoto signal   unvld noise
    always @(posedge div_clk or negedge sys_rst_n) begin
        if (sys_rst_n == 1'b0)
            noise_cnt <= 15'd0;
        else if( curr_st == CHECK_DATA_CODE && next_st == IDLE )
            noise_cnt <= 15'd0;
        else if ( noise_cnt < 15'd800)
            noise_cnt <= noise_cnt + 15'b1;
        else ;
    end
    
    
    always @(posedge div_clk or negedge sys_rst_n) begin
        if (sys_rst_n == 1'b0)
            remote_in_dly <= 1'b0;
        else
            remote_in_dly <= remote_in;
    end
    
    
    assign remote_pos =  remote_in & ~remote_in_dly ;
    assign remote_neg =  ~remote_in & remote_in_dly;
    
    
    // start cnt 1
    always @(posedge div_clk or negedge sys_rst_n  ) begin
        if (sys_rst_n == 1'b0)
            start_cnt1 <= 15'd0;
        else if( curr_st == CHECK_START_9MS && remote_in == 1'b0) begin
            if ( remote_neg == 1'b1 )
                start_cnt1 <= 15'd0;
            else
                start_cnt1 <= start_cnt1 + 15'b1;
        end
        else if( curr_st == CHECK_DATA_CODE)
            start_cnt1 <= 15'd0;
        else ;
    end
```

```verilog
// start cnt 2
always @(posedge div_clk or negedge sys_rst_n  ) begin
    if (sys_rst_n == 1'b0)
        start_cnt2 <= 15'd0;
    else if( curr_st == CHECK_START_4MS && remote_in == 1'b1) begin
        if ( remote_pos == 1'b1 )
            start_cnt2 <= 15'd0;
        else
            start_cnt2 <= start_cnt2 + 15'b1;
    end
    else if( curr_st == CHECK_START_9MS )
        start_cnt2 <= 15'd0;
    else ;
end


// CHECK_USER_CODE cnt
always @(posedge div_clk or negedge sys_rst_n  ) begin
    if (sys_rst_n == 1'b0)
        user_cnt <= 15'd0;
    else if( curr_st == CHECK_USER_CODE ) begin
        if ( remote_pos == 1'b1 )
            user_cnt <= user_cnt + 15'b1;
        else ;
    end
    else if( curr_st == CHECK_DATA_CODE)
        user_cnt <= 15'd0;
    else ;
end


// CHECK_DATA_CODE cnt
always @(posedge div_clk or negedge sys_rst_n) begin
    if (sys_rst_n == 1'b0)
        data_cnt <= 15'd0;
    else if( curr_st == CHECK_DATA_CODE ) begin
        if ( remote_pos == 1'b1 )
            data_cnt <= data_cnt + 15'b1;
        else ;
    end
    else if( curr_st == IDLE )
```

```verilog
            data_cnt <= 15'd0;
        else ;
    end

    always @(posedge div_clk or negedge sys_rst_n) begin
        if (sys_rst_n == 1'b0)
            data_judge_cnt <= 15'd0;
        else if( curr_st == CHECK_DATA_CODE ) begin
            if ( remote_pos == 1'b1 )
                data_judge_cnt <= 15'd0;
            else if ( remote_in == 1'b1 )
        data_judge_cnt <= data_judge_cnt + 15'b1;
        else ;
        end
        else ;
    end

    always @(posedge div_clk or negedge sys_rst_n  ) begin
        if (sys_rst_n == 1'b0)
            data_buf <= 16'd0;
        else if( curr_st == CHECK_USER_CODE && user_cnt == 15 )
            data_buf <= 16'd0;
        else if( curr_st == CHECK_DATA_CODE   ) begin
            if ( remote_neg == 1'b1 && data_judge_cnt > 10 )
                data_buf <= { data_buf[14:0], 1'b1 };
            else if ( remote_neg == 1'b1  )
                data_buf <= { data_buf[14:0], 1'b0 };
            else ;
        end
        else ;
    end

    always @(posedge  div_clk or negedge sys_rst_n ) begin
            if(sys_rst_n == 1'b0)
                led <= 8'd0;
            else if ( data_cnt >= 16 && remote_in == 1'b0 ) // Lock data_buf data code
                led <= data_buf[15:8] ;
    else ;
    end
```

endmodule

3.15.5 实验现象

上电,Quartus 编译完成后,下载 sof 文件到板子上,然后按下红外遥控器的"—"按钮,会有三个 LED 灯点亮,如图 3.67 所示,说明程序运行正确!

图 3.67 红外遥控实验现象

3.16 SDRAM 控制器实验

3.16.1 SDRAM 简介

存储器的分类如图 3.68 所示。

1. SDRAM 的特点

(1) 随机存取

当存储器中的消息被读/写时,所需时间(读取数据获得的时间)与这段信息所在的位置无关。当读取或写入顺序访问(Sequential Access)存储设备中的信息时,所需时间与位置就会有关系。

(2) 易失性

当电源关闭时 RAM 不能保留数据。如需保存数据,就必须把它们写入一个长期存储设备中(如 Flash)。RAM 和 ROM 的最大区别在于 RAM 在断电后所保存的数据会自动消失,而 ROM 不会。

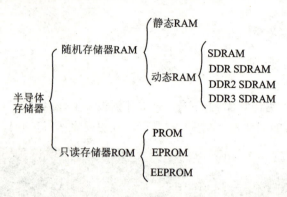

图 3.68 存储器的分类

(3) 需要刷新

动态随机存取存储器依赖内部存储区的电容器存储数据。电容未充电代表 0，充满电代表 1。由于电容器或多或少有漏电的情形，若不做特别处理，数据会渐渐随时间消失。刷新是指定期读取电容器的状态，然后按照原来的状态重新为电容器充电，弥补流失的电荷。需要不断刷新正好解释了随机存取存储器的易失性。

同步动态随机存储器(Synchronous Dynamic Random Access Memory)，中"同步"是指 Memory 工作需要同步时钟，内部的命令发送与数据传输都以它为基准；"动态"是指存储阵列需要不断地刷新来保证数据不丢失；随机是指数据不是线性依次存储，而是自由指定地址进行数据读/写。

图 3.69 是 HY57V641620 的实物图。

2. SDRAM 接口信号

① SD_CLK：时钟信号，为输入信号。SDRAM 所有输入信号的逻辑状态都需要通过 CLK 的上升沿采样确定。

② SD_CKE：时钟使能信号，为输入信号，高电平有效。CKE 信号的用途有两个：关闭时钟以进入省电模式；进入自刷新状态。CKE 无效时，SDRAM 内部所有与输入相关的功能模块停止工作。

③ SD_CS：片选信号，为输入信号，低电平有效。只有当片选信号有效后，SDRAM 才能识别控制器发送来的命令。设计时注意上拉。

④ SD_RAS：行地址选通信号，为输入信号，低电平有效。

⑤ SD_CAS：列地址选通信号，为输入信号，低电平有效。

⑥ SD_WE：写使能信号，为输入信号，低电平有效。

当然还包括 SD_BA[…]（即 bank）地址信号，这个需要根据不同的型号来确定，同样为输入信号；地址信号 SD_A[…]为输入信号；数据信号 SD_DQ[…]为输入/输出双向信号；数据掩码信号 DQM 为输入/输出双向信号，方向与数据流方向一致，高电平有效。当 DQM 有效时，数据总线上出现的对应数据字节被接收端屏蔽。

说明:

SD_CS/ SD_RAS/ SD_CAS/ SD_WE 组合完成 SDRAM 的各个命令,包括 ACT/WRITE 等。

3. SDRAM 的相关参数

(1) 突发长度(BL)

突发(burst)是指在同一行中相邻的存储单元连续进行数据传输的方式,连续传输的存储单元数量就是突发长度。

只要指定起始列地址与突发长度,内存就会依次自动对后面相应数量的存储单元进行读/写操作而不再需要控制器连续地提供列地址。BL 越长,对连续的大数据量传输就越有好处,但是对零散的数据,BL 太长反而会造成总线周期的浪费。但对于 DDR 而言,由于采用了预取技术,突发长度不再指所连续寻址的存储单元数量,而是指连续的传输周期数。

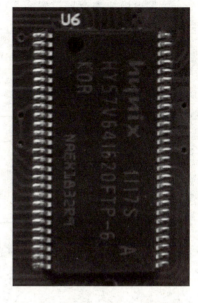

图 3.69 HY57V641620 实物图

(2) CL(CAS Latency)

CL 为内存存取数据所需的延迟时间,简单说就是内存接到 CPU 指令后的反应速度。作为衡量内存品质的重要指标,CL 延迟越小越好。

(3) t_{RCD}

t_{RCD} 指 RAS 至 CAS 延迟。RAS(数据请求后首先被激发)和 CAS(RAS 完成后被激发)并不是连续的,存在着延迟。

(4) t_{RAS}

t_{RAS} 指行有效至行预充电时间(active to precharge delay)。

(5) t_{RP}

t_{RP} 指行预充电时间(RAS precharge time)。也就是内存从结束一个行访问到重新开始的间隔时间。

4. SDRAM 的初始化(如图 3.70 所示)

① VDD(供输入缓冲器和逻辑电路)和 VDDQ(供输出缓冲器)上电,此期间 CKE 保持低电平。

② 开始发送时钟信号并使时钟使能信号 CKE 拉高。

③ 电源、时钟都稳定后,再等待 200 μs。

④ 发出预充电命令(PALL)。

⑤ 发出多个(8 个以上)刷新命令(REF)。

⑥ 发出模式寄存器设置命令(MRS),初始化模式寄存器(DDR2 中还有 EMRS,进行 ODT、OCD 等功能的设置和调整)。

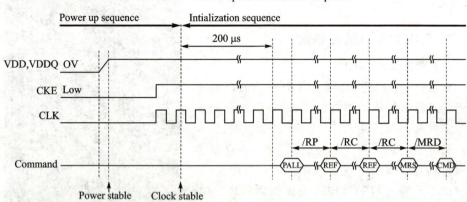

图 3.70　SDRAM 初始化图

5. SDRAM 的状态

Idle:空闲状态,是所有命令开始时的状态。

Row active:行地址有效,选定了操作对象的行地址和 BANK 地址,打开一个页面。

Precharge:预充电,当前行操作结束后要开始对一个新的行进行操作,必须先进行预充电操作。预充电后自动回到空闲状态。

Read and write:对操作对象执行相应的读/写操作,操作完成后自动回到行地址有效状态。

Read and write with auto precharge:对操作对象执行相应的读/写操作,操作完后自动进行预充电状态。

SDRAM 典型读时序(见图 3.71):ACT 指令先发出,然后发出 READ 指令,然后等待数据 Q1/Q2/Q3/Q4 返回。

SDRAM 典型写时序(见图 3.72):ACT 指令先发出,然后发出 WRITE 指令,同时送入 D1/D2/D3/D4 数据。

3.16.2　实验任务

将 SDRAM 初始化成功,然后写入一个数据,读出,比较相等后,点亮 4 个 LED 灯。

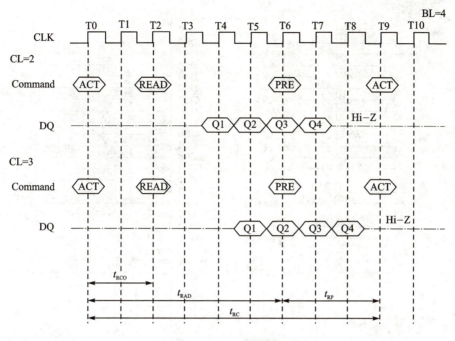

图 3.71　SDRAM 典型读时序

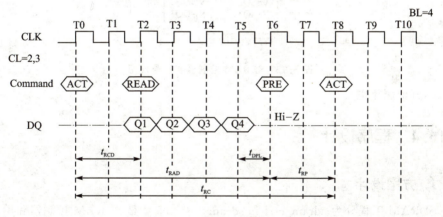

图 3.72　SDRAM 典型写时序

3.16.3　硬件设计

图 3.73 所示为 SDRAM 控制器实验原理图。

该 SDRAM 使用 16 bit 接口，容量大小为 64 Mbit，可以作为数据和程序的存储器。Qsys/Nios II 系统也可以使用 SDRAM 作为程序存储器。SDRAM 信号包括地址、数据、片选 CS、行选通 RAS、列选通 CAS、写使能 WE、时钟 SCLK 和时钟片选

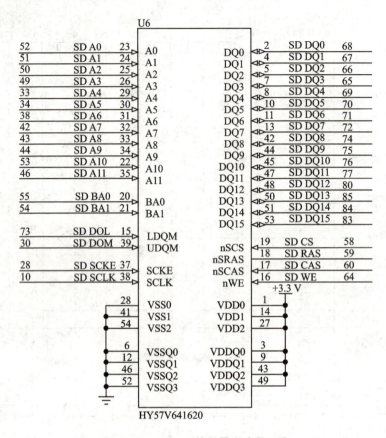

图 3.73 SDRAM 控制器实验原理图

SCKE,其他属于电源信号。

3.16.4 程序设计

1. 方案设计

SDRAM 工程分为 sdram_ctrl 和 sdram_test。前者是 SDRAM 控制器和用户侧接口,具有用户接口和 SDRAM 硬件接口,后者是用户使用的测试部分,可以完成发送数据到 SDRAM 中,再从 SDRAM 中取得数据。

图 3.74 所示是 sdram_ctrl 的时序。

当我们进行写操作时,sdram_req 拉高,rh_wl 读/写标志位拉低,在 sdram_data_w 上给出要写入的数据,sdram_addr 线上给出地址。以上四步要同时完成。完成后,sdram_ctrl 会发出 sdram_ack 结束信号。

当进行读操作时,rh_wl 拉高,sdram_req 拉高,给 addr 发送地址。以上三步同时完成。然后 sdram_ctrl 会将 sdram 中对应地址的数据发送出来,并且给出 sdram_

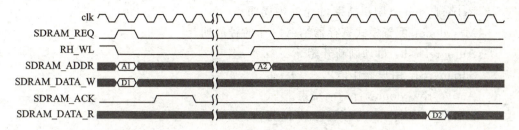

图 3.74 sdram_ctrl 时序

ack。由于 sdram 的读取周期问题,所以数据会在 sdram_ack 信号之后数周期输出。

SDRAM_CTRL 控制器状态机如图 3.75 所示。

SDRAM 控制器需要先进行初始化,然后进行模式设置,在模式设置里面,根据 SDRAM 器件手册要求进行设置即可。具体参数请参考代码。初始化完成后就可以进行读/写操作,在读或者写之前,需要先对 SDRAM 即将操作的行进行激活操作,然后才能发起读/写命令。

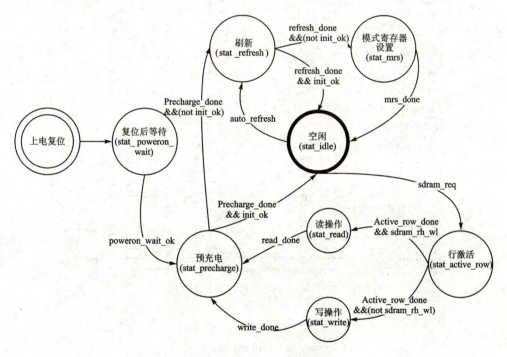

图 3.75 SDRAM 控制器状态机

2. 状态说明

Poweron_wait_ok:上电等待 200 μs 结束。

Precharge_done:预充电结束,1 个时钟周期。

Init_ok:初始化结束。

Refresh_done：刷新结束,1 个时钟周期。

Mrs_done：配置模式寄存器结束,3 个时钟周期。

Auto_refresh：自动刷新开始。

Sdram_req：用户操作请求。

Active_row_done：行激活结束。

Sdram_rh_wl：用户读/写控制,0：写；1：读。

Read_done：用户读操作结束,4 个时钟周期。

Write_done：用户写操作结束。

3. 预充电时间

在美光公司的手册里面,如图 3.76 所示,可以看出一个频率为 50 MHz(1/50 MHz=20 ns)的时钟即可完成 t_{RP}（即预充电时间 t_{RP}）的时间。

AC Characteristics Parameter		Symbol	-6 Min	-6 Max	-7E Min	-7E Max	-75 Min	-75 Max	Units	Notes
Access time from CLK (positive edge)	CL = 3	$t_{AC(3)}$	–	5.5	–	5.4	–	5.4	ns	27
	CL = 2	$t_{AC(2)}$	–	–	–	5.4	–	6	ns	
Address hold time		t_{AH}	1	–	0.8	–	0.8	–	ns	
Address setup time		t_{AS}	1.5	–	1.5	–	1.5	–	ns	
CLK high-level width		t_{CH}	2.5	–	2.5	–	2.5	–	ns	
CLK low-level width		t_{CL}	2.5	–	2.5	–	2.5	–	ns	
Clock cycle time	CL = 3	$t_{CK(3)}$	6	–	7	–	7.5	–	ns	23
	CL = 2	$t_{CK(2)}$	–	–	7.5	–	10	–	ns	23
CKE hold time		t_{CKH}	1	–	0.8	–	0.8	–	ns	
CKE setup time		t_{CKS}	1.5	–	1.5	–	1.5	–	ns	
CS#, RAS#, CAS#, WE#, DQM hold time		t_{CMH}	1	–	0.8	–	0.8	–	ns	
CS#, RAS#, CAS#, WE#, DQM setup time		t_{CMS}	1.5	–	1.5	–	1.5	–	ns	
Data-in hold time		t_{DH}	1	–	0.8	–	0.8	–	ns	
Data-in setup time		t_{DS}	1.5	–	1.5	–	1.5	–	ns	
Data-out High-Z time	CL = 3	$t_{HZ(3)}$	–	5.5	–	5.4	–	5.4	ns	10
	CL = 2	$t_{HZ(2)}$	–	–	–	5.4	–	6	ns	10
Data-out Low-Z time		t_{LZ}	1	–	1	–	1	–	ns	
Data-out hold time (load)		t_{OH}	2	–	3	–	3	–	ns	
Data-out hold time (no load)		t_{OHN}	1.8	–	1.8	–	1.8	–	ns	28
ACTIVE-to-PRECHARGE command		t_{RAS}	42	120,000	37	120,000	44	120,000	ns	
ACTIVE-to-ACTIVE command period		t_{RC}	60	–	60	–	66	–	ns	
ACTIVE-to-READ or WRITE delay		t_{RCD}	18	–	15	–	20	–	ns	
Refresh period (4,096 rows)		t_{REF}	–	64	–	64	–	64	ms	
Refresh period–Automotive (4,096 rows)		t_{REFAT}	–	16	–	16	–	16	ms	
AUTO REFRESH period		t_{RFC}	60	–	66	–	66	–	ns	
PRECHARGE command period		t_{RP}	18	–	15	–	20	–	ns	
ACTIVE bank a to ACTIVE bank b command		t_{RRD}	12	–	14	–	15	–	ns	
Transition time		t_T	0.3	1.2	0.3	1.2	0.3	1.2	ns	7
WRITE recovery time		t_{WR}	1 CLK + 6ns	–	1 CLK + 7ns	–	1 CLK + 7.5ns	–	–	24
			12	–	14	–	15	–	ns	25

图 3.76　SDRAM 器件参数

4. 刷新时间

根据美光公司推荐的初始化时序,预充电后需要进行刷新(见图 3.77),需要 t_{RFC} 时间,按照图 3.76,t_{RFC} 为 60 ns 左右,需要大于 60 ns(即 3 个 20 ns,时钟频率为 50 MHz 时,其周期为 20 ns)即可。

按照美光公司推荐的初始化时序,需要两次刷新。不过第二次刷新可以在模设置之后。

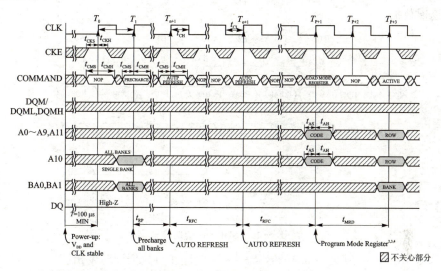

注：① 如果CS(CS信号为低电平，表示有效)信号在时钟上长为高电平，则所有的命令为无效命令。
② 模式寄存器可以在AUTO刷新命令加载之前进行配置。
③ JEDEC和PC100规范指定刷新时间为3个时钟周期。
④ 在命令发出之后，输出要保证是Z态(高阻态)。

图 3.77　SDRAM 刷新时间

美光公司手册推荐的 SDRAM 初始化要求，如图 3.78 所示。

The recommended power-up sequence for SDRAMs:
1. Simultaneously apply power to VDD and VDDQ.
2. Assert and hold CKE at a LVTTL logic LOW since all inputs and outputs are LVTTL-compatible.
3. Provide stable CLOCK signal. Stable clock is defined as a signal cycling within timing constraints specified for the clock pin.
4. Wait at least 100μs prior to issuing any command other than a COMMAND INHIBIT or NOP.
5. Starting at some point during this 100μs period, bring CKE HIGH. Continuing at least through the end of this period, 1 or more COMMAND INHIBIT or NOP commands must be applied.
6. Perform a PRECHARGE ALL command.
7. Wait at least ^{t}RP time; during this time NOPs or DESELECT commands must be given. All banks will complete their precharge, thereby placing the device in the all banks idle state.
8. Issue an AUTO REFRESH command.
9. Wait at least ^{t}RFC time, during which only NOPs or COMMAND INHIBIT commands are allowed.
10. Issue an AUTO REFRESH command.
11. Wait at least ^{t}RFC time, during which only NOPs or COMMAND INHIBIT commands are allowed.
12. The SDRAM is now ready for mode register programming. Because the mode register will power up in an unknown state, it should be loaded with desired bit values prior to applying any operational command. Using the LOAD MODE REGISTER command, program the mode register. The mode register is programmed via the MODE REGISTER SET command with BA1 = 0, BA0 = 0 and retains the stored information until it is programmed again or the device loses power. Not programming the mode register upon initialization will result in default settings which may not be desired. Outputs are guaranteed High-Z after the LOAD MODE REGISTER command is issued. Outputs should be High-Z already before the LOAD MODE REGISTER command is issued.
13. Wait at least ^{t}MRD time, during which only NOP or DESELECT commands are allowed.

At this point the DRAM is ready for any valid command.

图 3.78　SDRAM 初始化要求

CAS Latency 为 3 的 SDRAM 读时序(见图 3.79):送入读命令和读地址后,3 个周期返回数据。

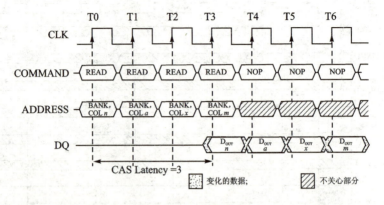

图 3.79 SDRAM 读时序

SDRAM 写时序(见图 3.80):写命令、写地址和写数据同时发出。

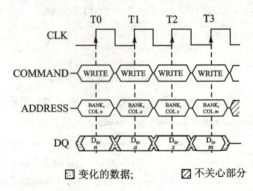

图 3.80 SDRAM 写时序(1)

图 3.81 是 SDRAM 一个写周期(T0~T1 的时间)内的各个接口信号的时序图,从图中可以看出 WE 为 0,代表低电平有效,写命令由 CS/RAS/CAS/WE(分别为 0 1 0 0)组合构成。

sdram_test 中,我们使用了三种状态来对 sd_ctrl 进行读/写:

① IDLE:上电时默认状态,在该状态时会使用定时器延时直到 sdram 初始化完成。

② WRITE:初始化完成后即转入此状态,在该状态时,只要有 wr_en 信号,就会向 sdram_ctrl 发送 req 信号,并且将 rh_wl 信号拉高,发送地址及数据。

③ READ:WRITE 状态时,延时计数到一定值(对 SDRAM 的读/写不能间隔过短),自动跳转到 READ 状态,在该状态里 rd_en 来时给出 req 和 addr,并且把 rw_wl 拉高,准备接收 data 数据。

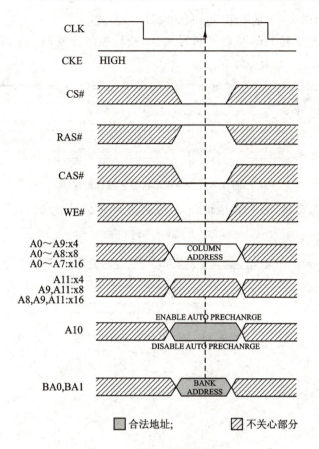

图 3.81 SDRAM 写时序(2)

sdram_test 内部状态跳转图如图 3.82 所示，包括 IDLE、WRITE 和 READ 三个状态。

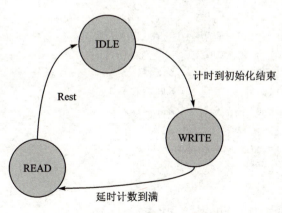

图 3.82 sdram_ctrl 状态跳转图

图 3.83 和图 3.84 是使用 Signaltap 抓取到的波形。

test 模块先发起 req、写数据和写地址,然后控制器返回 ack 即可完成一次写操作。

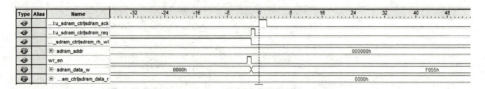

图 3.83 WRITE 状态下操作时序

test 模块先发起 req 和读地址,然后控制器返回 ack 和读数据。

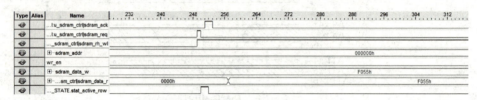

图 3.84 READ 状态下操作时序

源代码

```
module sdram_test(
    input           clk,
    input           reset_l,
    output          zs_ck,
    output          zs_cke,
    output          zs_cs_n,
    output          zs_ras_n,
    output          zs_cas_n,
    output          zs_we_n,
    output  [1:0]   zs_ba,
    output  [12:0]  zs_addr,
    output  [1:0]   zs_dqm,
    inout   [15:0]  zs_dq,
    output reg [7:0] led,
);

parameter   CHIP_ADDR_WIDTH = 13;
parameter   BANK_ADDR_WIDTH = 2;
parameter   ROW_WIDTH = 13;
```

```verilog
parameter    COL_WIDTH = 9;
parameter    DATA_WIDTH = 16;
parameter    CAS_LATENCY = 3'b011;


//three state
parameter    IDLE  = 3'b001;
parameter    WRITE = 3'b010;
parameter    READ  = 3'b100;

parameter    div_400us = 15'd25000;          //delay number,for initialization of sdram

parameter    sdram_addr_num = 24'b0;         //row col bank: 13 + 9 + 2

parameter    sdram_data_w_num = 16'b1111000001010101;         //f055

reg                     sdram_req;
reg     [23:0]          sdram_addr;
reg                     sdram_rh_wl;
reg     [15:0]          sdram_data_w;

reg     [2:0]           current_state;
reg     [2:0]           current_state_dly1;

reg     [2:0]           next_state;

reg     [14:0]          init_wait_cnt;
reg     [7:0]           wr_cnt;
reg                     sdram_data_r_en_1 ;
reg                     sdram_ack_1;
reg     [15:0]          sdram_data_r_lock ;

wire    [15:0]          sdram_data_r;
wire                    sdram_data_r_en;
wire                    sdram_ack;

wire                    rd_en;
wire                    wr_en;
```

```verilog
always@(posedge clk or negedge reset_1) begin
    if(! reset_1)
        current_state <= IDLE;
    else
        current_state <= next_state;
end

always@(posedge clk or negedge reset_1) begin
    if(! reset_1)
        init_wait_cnt <= 15'b0;
    else if( init_wait_cnt <= ( div_400us - 1 ))
        init_wait_cnt <= init_wait_cnt + 1;
    else ;
end

always@(posedge clk or negedge reset_1) begin
    if(! reset_1)
            wr_cnt <= 8'b0;
    else if( current_state == WRITE && wr_cnt < 8'd250  )
            wr_cnt <= wr_cnt + 8'b1;
    else ;
end

always@( * ) begin
    case(current_state)
        IDLE:begin
                if(init_wait_cnt >= (div_400us - 1))
                        next_state = WRITE;
                else
                        next_state = IDLE;
        end
        WRITE:begin
                if(wr_cnt == 8'd250)
                        next_state = READ;
                else
                        next_state = WRITE;
        end
        READ:
                next_state = READ;
```

```verilog
        endcase
end

always@(posedge clk or negedge reset_l) begin    //delay button's signal,
                                                 //make this stable
    if(! reset_l)
        current_state_dly1 <= 3'b0;
    else
        current_state_dly1 <= current_state;
end

assign wr_en = ( current_state == WRITE ) & ( current_state_dly1 != WRITE ) ;

assign rd_en = ( current_state == READ ) & ( current_state_dly1 != READ ) ;

always@(posedge clk or negedge reset_l) begin
    if(! reset_l) begin
        sdram_rh_wl <= 1'b1;
        sdram_req <= 1'b0;
        sdram_data_r_en_1 <= 1'b0;
        sdram_ack_1 <= 1'b0;
        sdram_addr <= 0;
        sdram_data_w <= 0;
    end
    else begin
    if ( current_state == WRITE) begin
        sdram_data_r_en_1 <= sdram_data_r_en;
        sdram_ack_1 <= sdram_ack;

        sdram_addr <= sdram_addr_num;
        sdram_data_w <= sdram_data_w_num;
        sdram_req <= wr_en;
        sdram_rh_wl <= 1'b0;
    end
    else if ( current_state == READ) begin
        sdram_rh_wl <= 1'b1;
        sdram_req <= rd_en;
        sdram_data_r_en_1 <= sdram_data_r_en;
        sdram_ack_1 <= sdram_ack;
```

```verilog
        end

    end
end

always@(posedge clk or negedge reset_l) begin
    if(! reset_l)
        sdram_data_r_lock <= 16'b0;
    else if ( sdram_ack == 1'b1 && sdram_rh_wl == 1'b1 )
        sdram_data_r_lock <= sdram_data_r;
    else ;
end

always@(posedge clk or negedge reset_l) begin
    if(! reset_l)
        led <= 8'b0;
    else if ( sdram_data_r_lock == sdram_data_w )
        led <= 8'h55;
    else ;
end

sdram_ctrl u_sdram_ctrl(

    .clk(clk),
    .reset_l(reset_l),

    .sdram_req              (sdram_req),
    .sdram_ack              (sdram_ack),
    .sdram_addr             (sdram_addr),
    .sdram_rh_wl            (sdram_rh_wl),
    .sdram_data_w           (sdram_data_w),
    .sdram_data_r           (sdram_data_r),
    .sdram_data_r_en        (sdram_data_r_en),

    .zs_ck                  (zs_ck),
    .zs_cke                 (zs_cke),
    .zs_cs_n                (zs_cs_n),
    .zs_ras_n               (zs_ras_n),
    .zs_cas_n               (zs_cas_n),
```

```verilog
        .zs_we_n        (zs_we_n),
        .zs_ba          (zs_ba),
        .zs_addr        (zs_addr),
        .zs_dqm         (zs_dqm),
        .zs_dq          (zs_dq)
    );


endmodule
module sdram_ctrl(
            //input
            clk,
            reset_l,
            //client interface
              sdram_req,
              sdram_ack,
              sdram_addr,
              sdram_rh_wl,
              sdram_data_w,
              sdram_data_r,
              sdram_data_r_en, //indicate read data valid
              //chip interface
              zs_ck,
            zs_cke,
            zs_cs_n,
            zs_ras_n,
            zs_cas_n,
            zs_we_n,
            zs_ba,
            zs_addr,
            zs_dqm,
            zs_dq
              );

parameter   CHIP_ADDR_WIDTH = 13;
parameter   BANK_ADDR_WIDTH = 2;
parameter   ROW_WIDTH = 13;
parameter   COL_WIDTH = 9;
parameter   DATA_WIDTH = 16;
```

```verilog
    parameter      CAS_LATENCY = 3'b011;
//auto refresh cycle calculate:
//each row must be refreshed every 64 ms, and just refresh one row each time.
//and this chip have 8 192 rows, so each refresh interval must less than 64 ms/8 192 = 7.8125 μs
//50 MHz clock should count to : 7.812 5×1 000/20 = 390,then have one time refresh
    parameter      AUTO_REFRESH_CYCLE   = 390;
    parameter      POWERON_WAIT_CYCLE   = 10 000;   //need wait 200 μs,based on 50m clock

    input     clk;
    input     reset_l;

    input     sdram_req;
    output    sdram_ack;
    input     [ROW_WIDTH + COL_WIDTH + BANK_ADDR_WIDTH - 1:0] sdram_addr;
    input     sdram_rh_wl;
    input     [DATA_WIDTH - 1:0]    sdram_data_w;
    output    [DATA_WIDTH - 1:0]    sdram_data_r;
    output                          sdram_data_r_en;

    output    zs_ck;
    output    zs_cke;
    output    zs_cs_n;
    output    zs_ras_n;
    output    zs_cas_n;
    output    zs_we_n;
    output    [BANK_ADDR_WIDTH - 1:0] zs_ba;
    output    [CHIP_ADDR_WIDTH - 1:0] zs_addr;
    output    [1:0]    zs_dqm;
    inout     [DATA_WIDTH - 1:0]    zs_dq;

    wire      zs_ck;
    assign    zs_ck = clk;
    wire      zs_cke;
    assign    zs_cke = 1'b1;
    wire      zs_cs_n;
    wire      zs_ras_n;
    wire      zs_cas_n;
    wire      zs_we_n;
    reg       [3:0]    sdram_cmd;
```

```verilog
reg     [BANK_ADDR_WIDTH-1:0]   zs_ba;
reg     [CHIP_ADDR_WIDTH-1:0]   zs_addr;
reg     [1:0]   zs_dqm;
reg     zs_dq_o_en;
reg     [DATA_WIDTH-1:0]    zs_dq_o;
wire    [DATA_WIDTH-1:0]    zs_dq_i;
reg     [DATA_WIDTH-1:0]    sdram_data_r;
reg     sdram_data_r_en;
reg     sdram_ack;

parameterstat_poweron_wait  = 8'b00000001;
parameterstat_precharge     = 8'b00000010;
parameterstat_refresh       = 8'b00000100;
parameterstat_mrs           = 8'b00001000;
parameterstat_idle          = 8'b00010000;
parameterstat_active_row    = 8'b00100000;
parameterstat_read          = 8'b01000000;
parameterstat_write         = 8'b10000000;
reg     [7:0]CUR_STATE;
reg     [7:0]NEXT_STATE;

reg             auto_refresh;
reg     [15:0]  auto_refresh_cnt;
reg             poweron_wait_ok;
reg             init_ok;
reg             precharge_done;
reg             refresh_done;
reg             mrs_done;
reg             active_row_done;
reg             read_done;
reg             write_done;

reg     [15:0]  poweron_wait_cnt;
reg     [3:0]   status_running_cnt;

assign zs_dq = (zs_dq_o_en == 1'b1)? zs_dq_o:{DATA_WIDTH{1'bz}};
assign zs_dq_i = zs_dq;
assign {zs_cs_n, zs_ras_n, zs_cas_n, zs_we_n} = sdram_cmd;
```

```verilog
always @ ( negedge reset_1 or posedge clk )
    begin
        if(reset_1 == 1'b0) begin
            CUR_STATE <= stat_poweron_wait ;
        end
        else begin
            CUR_STATE <= NEXT_STATE;
        end
    end
always @ ( * )
    begin
      NEXT_STATE <= stat_idle;
      case (CUR_STATE)
        stat_poweron_wait :begin
            if(poweron_wait_ok == 1'b1) begin
                NEXT_STATE <= stat_precharge;
            end
            else begin
                NEXT_STATE <= stat_poweron_wait;
            end
          end
        stat_precharge :  begin
            if(precharge_done == 1'b1) begin
                if(init_ok == 1'b1) begin
                    NEXT_STATE <= stat_idle;
                end
                else begin
                    NEXT_STATE <= stat_refresh;
                end
            end
            else begin
                NEXT_STATE <= stat_precharge;
            end
          end
        stat_refresh :  begin
            if(refresh_done == 1'b1) begin
                if(init_ok == 1'b1) begin
                    NEXT_STATE <= stat_idle;
                end
```

```verilog
        else begin
            NEXT_STATE <= stat_mrs;
          end
        end
      else begin
        NEXT_STATE <= stat_refresh;
      end
end
stat_mrs : begin
    if(mrs_done == 1'b1) begin
      NEXT_STATE <= stat_idle;
    end
    else begin
      NEXT_STATE <= stat_mrs;
    end
end
stat_idle :  begin
    if(auto_refresh == 1'b1) begin
      NEXT_STATE <= stat_refresh;
    end
    else if(sdram_req == 1'b1) begin
      NEXT_STATE <= stat_active_row;
    end
    else begin
      NEXT_STATE <= stat_idle;
    end
end
stat_active_row :  begin
    if(active_row_done == 1'b1) begin
        if(sdram_rh_wl == 1'b1) begin
          NEXT_STATE <= stat_read;
        end
        else begin
          NEXT_STATE <= stat_write;
        end
    end
    else begin
      NEXT_STATE <= stat_active_row;
    end
end
stat_read ;begin
    if(read_done == 1'b1) begin
        NEXT_STATE <= stat_precharge;
    end
    else begin
        NEXT_STATE <= stat_read;
```

```verilog
                end
            end
            stat_write :begin
                if(write_done == 1'b1) begin
                    NEXT_STATE <= stat_precharge;
                end
                else begin
                    NEXT_STATE <= stat_write;
                end
            end
            default : begin
                NEXT_STATE <= stat_idle ;
            end
        endcase
end

//sdram acknology control
always @ ( negedge reset_l or posedge clk )
    begin
        if(reset_l == 1'b0) begin
            sdram_ack <= 1'b0;
        end
        else begin
          sdram_ack <= 1'b0;
          if(CUR_STATE == stat_active_row) begin
              sdram_ack <= 1'b1;
          end
          else if(sdram_req == 1'b1) begin
              sdram_ack <= 1'b0;
          end
        end
    end

//stat_poweron_wait
always @ ( negedge reset_l or posedge clk )
    begin
        if(reset_l == 1'b0) begin
            poweron_wait_cnt <= 16'b0;
            poweron_wait_ok  <= 1'b0;
        end
        else begin
          poweron_wait_ok <= 1'b0;
          if(CUR_STATE == stat_poweron_wait) begin
              if(poweron_wait_cnt >= POWERON_WAIT_CYCLE) begin
                  poweron_wait_ok <= 1'b1;
              end
              else begin
```

```verilog
                    poweron_wait_cnt <= poweron_wait_cnt + 1;
                end
            end
            else begin
                poweron_wait_cnt <= 16'b0;
            end
        end
    end
//auto refresh control
always @ ( negedge reset_l or posedge clk )
    begin
        if(reset_l == 1'b0) begin
            auto_refresh_cnt <= 16'b0;
            auto_refresh <= 1'b0;
        end
        else begin
          if(auto_refresh == 1'b0) begin
            auto_refresh_cnt <= auto_refresh_cnt + 1;
          end
          else begin
            auto_refresh_cnt <= 16'b0;
          end
          if(auto_refresh_cnt >= AUTO_REFRESH_CYCLE) begin
            auto_refresh <= 1'b1;
          end
          else if(CUR_STATE == stat_refresh) begin
                auto_refresh <= 1'b0;
          end
        end
    end

//status running control
always @ ( negedge reset_l or posedge clk )
    begin
        if(reset_l == 1'b0) begin
            status_running_cnt <= 4'b0;
        end
        else begin
          if(precharge_done || refresh_done || mrs_done || active_row_done || read_done || write_done) begin
                status_running_cnt <= 4'b0;
          end
          else if(CUR_STATE == stat_precharge || CUR_STATE == stat_refresh ||
                  CUR_STATE == stat_mrs|| CUR_STATE == stat_active_row ||
                  CUR_STATE == stat_read || CUR_STATE == stat_write) begin
                status_running_cnt <= status_running_cnt + 4'b1;
          end
```

```verilog
            else begin
                status_running_cnt <= 4'b0;
            end
        end
    end

//other status control
always @ ( negedge reset_l or posedge clk )
    begin
        if(reset_l == 1'b0) begin
            sdram_cmd <= {4{1'b1}};
            zs_ba <= {BANK_ADDR_WIDTH{1'b0}};
            zs_addr <= {CHIP_ADDR_WIDTH{1'b0}};
            zs_dqm <= 2'b0;
            zs_dq_o_en <= 1'b0;
            zs_dq_o <= {DATA_WIDTH{1'b0}};

            init_ok <= 1'b0;
            precharge_done <= 1'b0;
            refresh_done <= 1'b0;
            mrs_done <= 1'b0;
            active_row_done <= 1'b0;
            read_done <= 1'b0;
            write_done <= 1'b0;

            sdram_data_r_en <= 1'b0;
            sdram_data_r <= {DATA_WIDTH{1'b0}};
        end
        else begin
            precharge_done <= 1'b0;
            refresh_done <= 1'b0;
            mrs_done <= 1'b0;
            active_row_done <= 1'b0;
            read_done <= 1'b0;
            write_done <= 1'b0;
            zs_ba <= sdram_addr[ROW_WIDTH + COL_WIDTH + BANK_ADDR_WIDTH - 1:ROW_WIDTH + COL_WIDTH];
            zs_dq_o_en <= 1'b0;
            sdram_data_r_en <= 1'b0;
        case (CUR_STATE)
        stat_precharge :begin
            sdram_cmd <= 4'b0010;
            zs_addr[10] <= 1'b1;
            precharge_done <= 1'b1;
        end
        stat_refresh :begin
            if(status_running_cnt == 4'b0) begin
```

```verilog
                sdram_cmd <= 4'b0001;
            end
        else begin
            sdram_cmd <= 4'b0111;//none operating mode
        end
        if(status_running_cnt >= 4'd8) begin
            refresh_done <= 1'b1;
        end
    end
stat_mrs :begin
        if(status_running_cnt == 4'b0) begin
            sdram_cmd <= 4'b0000;
            zs_addr <= {{3{1'b0}},1'b0,2'b00,CAS_LATENCY,4'h0};
        end
        else begin
            sdram_cmd <= 4'b0111;//none operating mode
        end
        if(status_running_cnt >= 4'd3) begin
            mrs_done <= 1'b1;
            init_ok <= 1'b1;
        end
    end
stat_active_row :begin
        sdram_cmd <= 4'b0011;
            zs_addr <= sdram_addr[ROW_WIDTH + COL_WIDTH - 1:COL_WIDTH];
            active_row_done <= 1'b1;
    end
stat_read :begin
        if(status_running_cnt == 4'd0) begin
            sdram_cmd <= 4'b0101;
            zs_addr <= sdram_addr[COL_WIDTH - 1:0];
        end
        if(status_running_cnt == 4'd3) begin
            read_done <= 1'b1;
            sdram_data_r_en <= 1'b1;
            sdram_data_r <= zs_dq_i;
        end
    end
stat_write :begin
        zs_dq_o_en <= 1'b1;
        if(status_running_cnt == 4'd0) begin
            sdram_cmd <= 4'b0100;
            zs_addr <= sdram_addr[COL_WIDTH - 1:0];
            zs_dq_o <= sdram_data_w;
        end
        if(status_running_cnt == 4'd1) begin
            write_done <= 1'b1;
```

```
                end
            end
            stat_idle :begin
                sdram_cmd <= 4'b1111; //command disable
                zs_addr <= {CHIP_ADDR_WIDTH{1'b0}};
            end
            default : ;
        endcase
    end
end

endmodule
```

3.16.5 实验现象

程序编译后,下载到开发板里面,会有 4 个 LED 灯亮(见图 3.85),说明 SDRAM 读/写成功。

图 3.85　SDRAM 控制器实验读/写实验现象

第 4 章
设计思想和感悟

设计思想主要是有经验的老员工的经验积累,新员工和初学者最欠缺的就是设计思想,因为这个是需要经过大型项目锤炼才能获得的。

可能很多初学者不理解为什么要有规范,为什么"大牛"和"菜鸟"写的代码差异很大,或者"菜鸟"的 Bug 很多。作为过来人,我建议"菜鸟"们先不要纠结对错,先按照"大牛"的规范和思路学习和实践,等"菜鸟"飞起来的那天,就会觉得之前的规范和思想有多么重要,那些都是前人的精华总结。

下面几点是笔者做过多个复杂项目后的几个思想总结和感悟。

4.1 代码简单化

很多初学者和没有经过正规项目实践的人可能会觉得代码写的越复杂越有水平,笔者最开始学习 Verilog 也是这样想的,每每看到复杂的代码都崇拜不已。

其实不是这样的。当前项目越来越复杂,新写的代码都需要经过多轮检视和后续项目重用,如果您的代码写的复杂无比,那么带来的检视、重用工作量就是巨大的,也是项目周期不容许的,所以最好是使用简单的语句和逻辑实现复杂的功能,越简单越好。能把一个复杂的东西使用简单的语句表达出来,是一种很好的能力。

代码简单化有几点建议:
① 使用最基础和常见的语法,也有利于工具的资源优化。
② 少用复杂的语法,如"~^""==="等。
③ 复杂功能分析清楚,语句之间越少耦合越好。
④ 尽量不要使用状态机,状态机综合的电路较难分析,而且与其他电路相比,还需要分析状态机覆盖率。

4.2　注释层次化

代码越复杂,注释的意义越大,注释分层可以提高对代码的理解和方案的划分。注释层次化有几点建议:
① 使用步骤 1、2、3、4 等编号描述.V 文件的结构划分和功能。
② 每个步骤里面可以包括子步骤 1、2、3、4 等。
③ 每个子步骤如果还是很复杂,可以继续划分子子步骤。
④ 每个子步骤包括几个 always 或者 assign 块逻辑。

4.3　交互界面清晰化

交互界面清晰化指的是模块间接口或者模块内子模块交互简单、清晰,没有歧义。因为涉及到接口配合的地方一般都不是一个人完成的代码,凡不是一个人完成的事情都需要理解对方的意图,都是容易出问题的地方。

交互界面清晰化有几点建议:
① 接口定义越简单越好,需要避免太多耦合。
② 信号功能描述清晰,没有歧义。
③ 信号最好在任何时候代表的都是相同的功能。
④ 从系统方案层面划分接口,保证接口耦合性最少

4.4　模块划分最优化

复杂系统一般都是由多个模块一起完成的,如果模块划分不合理,会出现模块间耦合太多,导致系统不够健壮和清晰,很容易出问题,而且系统接受新需求能力不强,后续项目重用也比较麻烦。所以模块划分是非常重要的一个环节。让模块划分最优化是一个系统工程师必备的技能。

模块划分最优化有几点建议:
① 完全理解系统的需求和规格。
② 对系统的实现要做到非常清楚。
③ 划分模块需要和熟悉系统的人讨论,大家都觉得是最优的时候才行。
④ 模块间接口简单、清晰,没有歧义。
⑤ 模块功能比较完整,同一个功能最好使用一个模块实现。
⑥ 系统划分也要方便验证。

4.5 方案精细化

设计方案是编写代码的前提,复杂些的项目如果没有方案,根本没有办法写出代码。另外,很多初学者都希望多些时间写代码,其他时间能少则少,其实这种思想是错误的,磨刀不误砍柴工,有了详细的设计方案后,写代码可以说是体力活。

笔者之前有一个项目,设计方案的时间是 3 个月,写代码的时间是 1.5 个月,方案时间远大于编码时间。另外,方案必须达到能指导编码的阶段才能进行编码,其中包括:功能描述、架构实现、表现描述、模块划分、模块实现细节和时序描述、接口描述等。

方案精细化有几点建议:
① 设计方案细化到 always 块层次。
② 模块实现细节描述清楚。
③ 重点电路和复杂电路的时序需要在编码前画出。
④ 设计方案需要经过专家评审。
⑤ 设计方案需要有架构和模块划分描述。
⑥ 设计方案需要包括资源占用的评估报告。

4.6 时序流水化

时序设计是代码编写的前提,流水处理包括两种。第一种是通过时隙划分的方式实现流水,第二种是通过寄存器延时的方式进行流水设计。

第一种方式

经过几个复杂项目,我们发现,如果时序是时隙流水的形式,可以极大地减小设计的复杂度。比如操作一个 RAM,可以使用四个时隙去划分,第一个时隙是逻辑读/写,第二个时隙是 CPU 读/写,那么这种就不需要做读/写冲突处理。四个时钟周期作为一个流水线不停地流水执行,这样设计简化很多。流水时隙设计者有个前提就是流水线能满足性能要求,比如需要每个周期都读/写 RAM,那么用四个时隙周期去划分,第一个时隙是逻辑读/写,第二个时隙是 CPU 读/写就不可行了,这个时候需要做 RAM 读/写冲突处理。

第二种方式

还有一种是非时隙流水处理,这种是固定使用寄存器延时进行处理,比如从 FIFO 读出数据的第一个时钟周期进行数据校验,第二个时钟周期进行数据处理,第三个时钟周期进行数据拼接,第四个时钟周期进行数据的 CRC 校验。这种处理方式也是流水处理的一种,其性能很高,也是非常常见的设计方法之一。

时序流水化有几点建议：

① 满足性能要求的前提时隙化处理。

② 通过寄存器延时的方式难度大于时隙化处理方式。

③ 每个时隙或者延迟周期处理事情要明确、单一。

参考文献

[1] 长沙太阳人电子有限公司. SMC1602A LCM 手册. 长沙太阳人电子有限公司,2012.
[2] Micron Technology,Inc. 64 M SDRAM 颗粒手册. Micron Technology,Inc,2009.
[3] 恒创科技工作室. 恒创科技暴风四代开发板使用手册. 恒创科技工作室,2014.
[4] 恒创科技工作室. 恒创科技暴风四代开发板配套程序. 恒创科技工作室,2014.
[5] SECONS Ltd. 640×480 VGA 时序标准. [2016-09-01]. http://tinyvga.com/vga-timing/640x480@60Hz.